CU00926407

Racial Science and British Society, 1930–62

Racial Science and British Society, 1930–62

Gavin Schaffer

© Gavin Schaffer 2008

All rights reserved. No reproduction, copy or transmission of this publication may be made without written permission.

No portion of this publication may be reproduced, copied or transmitted save with written permission or in accordance with the provisions of the Copyright, Designs and Patents Act 1988, or under the terms of any licence permitting limited copying issued by the Copyright Licensing Agency, Saffron House, 6-10 Kirby Street, London EC1N 8TS.

Any person who does any unauthorized act in relation to this publication may be liable to criminal prosecution and civil claims for damages.

The author has asserted his right to be identified
as the author of this work in accordance with the Copyright, Designs and Patents Act 1988.

First published 2008 by
PALGRAVE MACMILLAN

Palgrave Macmillan in the UK is an imprint of Macmillan Publishers Limited, registered in England, company number 785998, of Houndmills, Basingstoke, Hampshire RG21 6XS.

Palgrave Macmillan in the US is a division of St Martin's Press LLC, 175 Fifth Avenue, New York, NY 10010.

Palgrave Macmillan is the global academic imprint of the above companies and has companies and representatives throughout the world.

Palgrave® and Macmillan® are registered trademarks in the United States, the United Kingdom, Europe and other countries.

ISBN-13: 978–0–230–00892–2   hardback
ISBN-10: 0–230–00892–5   hardback

This book is printed on paper suitable for recycling and made from fully managed and sustained forest sources. Logging, pulping and manufacturing processes are expected to conform to the environmental regulations of the country of origin.

A catalogue record for this book is available from the British Library.

Library of Congress Cataloging-in-Publication Data

Schaffer, Gavin, 1976–
  Racial science and British society, 1930–1962 / Gavin Schaffer.
     p. cm.
  Includes bibliographical references and index.
  ISBN-10: 0–230–00892–5 (alk. paper)
  ISBN-13: 978–0–230–00892–2
     1. Great Britain–Race relations–History–20th century. 2. Racism–Great Britain–History–20th century. 3. Great Britain–Politics and government–20th century. 4. World War, 1939–1945–Social aspects–Great Britain. 5. Great Britain–Social policy. I. Title.
DA125.A1S33 2008
305.800941′09041–dc22   2008016150

10 9 8 7 6 5 4 3 2 1
17 16 15 14 13 12 11 10 09 08

Printed and bound in Great Britain by
CPI Antony Rowe, Chippenham and Eastbourne

To C.H.T.
with love

Contents

Acknowledgements		ix
1	**Introduction**	**1**
	The end of the race concept	1
	Science or society? Who takes the lead?	3
	The 'post 1945' thesis	4
	The meanings of race	6
	Deconstructing racial science and British society	8
	Race and immigration policy in early twentieth-century Britain	11
2	**Rethinking Interwar Racial Reform: the 1930s**	**15**
	Four racial field studies	16
	Changing methods and changing politics	25
	The limits of racial reform before the Second World War	39
	Science and society in the 1930s	48
	Refuge, race and restriction: Britain and the Jewish refugees from Nazism	53
3	**The Challenge of War: the 1940s**	**63**
	Scientists and the wartime racial agenda	63
	Scientists versus the state? Enemy aliens and Britain's black subjects	79
	The perilous mission of the believers: protecting the idea of race in the face of Nazism and the Holocaust	96
	Sharing the biologists' uncertainty? The European Volunteer Workers policy after the Second World War	107
4	**Race on the Retreat? The 1950s and 1960s**	**115**
	Stepping back from society: the progressives, Lysenko and Soviet science	115
	British biologists and the UNESCO statements on race	120
	Progressives and conservatives in the 1950s: a case of reunification?	132
	British science and post-war Commonwealth immigration to Britain	148
5	**Epilogue**	**166**

Notes	172
Select Bibliography	216
Index	229

Acknowledgements

I would like to thank the Arts and Humanities Research Council, the British Academy, and the Centre for European and International Studies Research at the University of Portsmouth for supporting various stages of this project. I am also indebted to a number of colleagues and friends who have helped and advised me along the way: Gemma Romain, Dan Stone, John and Marilyn Lehmann, Bill Tucker, Barry Mehler, Tom Lawson, Michael Banton, Matthew Taylor, Keith Bearpark, Wendy Ugolini, Colin Holmes, Panikos Panayi, Lesley Hall and Tim Grady. Most of all, I would like to thank Tony Kushner for his guidance and friendship over the last seven years. I am grateful for the support of my colleagues at the University of Portsmouth, in particular Brad Beaven, Bob Kiehl, June Purvis, Dave Andress, Sue Bruley, Graham Attenborough, Monica Riera, Martin Evans, Rob James and James Thomas. This project would not have been possible without the assistance of some very dedicated and knowledgeable archivists. Special mention must be given to Phil Montgomery and Lisa Moellering from the Woodson Research Centre at Rice University and all the staff at the Liddell Hart archive in King's College and the Wellcome Archive. Additionally, I would like to thank the Galton Institute for allowing me to have access to the records of the Eugenics Society. Finally, I would like to thank my parents Jo and Jeff Schaffer for their love and unstinting support.

1
Introduction

The end of the race concept

In 1949 the historian of science Charles Singer wrote to his friend, the biologist and zoologist Julian Huxley. In jovial form, Singer informed Huxley that he had come across 'a superb use of the word <u>race</u> which will stick in your memory'. Singer had found this 'beautiful specimen' in the *Dictionary of National Biography*'s obituary of a certain Lord Acton. It read: 'Acton, who though <u>a Roman Catholic by race and training</u>, was deeply hostile to the arbitrary power of the Pope, owed his existence to a papal dispensation'.[1] Singer seemingly felt that Huxley would find this amusing and probably he did. The two men had sporadically worked together on various publications in the preceding 15 years which had tried to pin down the meaning of race.[2] In the context of their shared interest, the bizarre if benign notion that one could be a Catholic 'by race' would have appealed to them both.

Huxley and Singer would no doubt have found it less amusing that in the 60 years which have followed this letter the idea of race and the appropriateness of its usage have hardly been clarified. In fact, the use of the term in our society, whilst certainly different, remains both diverse and unpredictable.[3] In what has been a long post-war battle against the injustices of prejudice and discrimination, the notion of race has survived. It has done so not only through the agency of those who have wished to keep it alive but also, perhaps more significantly, by hanging onto the coat-tails of the very language which has been employed in British society to mitigate and ameliorate its effects, in the discourses of anti-racism and race relations. The unsatisfactory nature of this state of affairs has not been lost on scholars of contemporary and historical racism. Robert Miles, for example, has called for an end to the use of racial language of any kind, believing that such vocabulary breathes life into what should now be regarded as a long discredited idea. The task for scholars, Miles contends, should be to make:

> ...the phenomenal categories of 'race' and 'race relations' the object of a critique. That is to say, we must not use these categories analytically, but rather we should ask how and why they came to be constructed to give apparent sense to the phenomenal world.[4]

This criticism of the survival of race can be read as an intellectual objection to the continuing presence in our society of a term which is no longer given to be scientifically credible. In Bob Carter's words: 'there is no theoretical definition of race...whatever meanings the term race has are those given it in use by lay actors'.[5] However, most criticism of the lingering idea of race is not solely intellectual. After all, there are thousands of uncontroversial words and ideas in the English lexicon which do not accurately reflect any scientific reality. In a society which concerns itself little over the facts that we do not have lead in our pencils and that the sky is not actually blue (or for that matter that black and white people are respectively neither black nor white), the campaign against the continuation of 'race' must be set in parameters beyond the idea of a lay/scientific disparity of meaning.

Miles's claims about race are scientifically meaningful, but his agenda is political and moral. Writing retrospectively about the methodological issues which surrounded early sociological studies of race in the 1950s, Michael Banton highlighted the problem that 'a scientific error of the mid-nineteenth century had been incorporated into the ordinary language and had moulded the way everyone thought of ethnic and racial matters'.[6] Here, in a nutshell, lies the root of Miles's objection to the ongoing use of the term 'race'. If it were a value-neutral idea, or an uncontroversial one, then its continuation in lay or political language would not matter. It does though matter, because, as Banton explains, it has 'moulded the way' that people think.[7] Miles's agenda has thus been to end the use of the term, not for the sake of science, but as a means to shut the door on ideas which have legitimised a century of international conflict, genocide and discrimination; a way to undermine a concept that Gilroy has described as 'modernity's most pernicious signature'.[8]

In British science, this anti-racial agenda has been long in the making. Huxley and Haddon's seminal 1935 text *We Europeans*, like Miles 50 years later, also called for the end of the use of the term 'race', as the authors attempted to refute the scientific legitimacy of extreme racial theorising, with a clear eye on Nazi Germany.⁹ The point here is not that the *We Europeans* scholars shared Miles's attitude to race or understood the concept in the same way. As we shall see below they certainly did not. It is merely to stress the length of time during which British experts have tried to engage with the public's understanding of race and influence it. This attempt at influence has never maintained a coherent direction. Just as the scholars cited above have written in favour of the end of race, so others have written in favour of preserving the idea.

Science or society? Who takes the lead?

The passage of thinking between experts and society on racial matters should not be seen as one-way traffic. It has often been factors outside science that have driven scholars to write on race, creating an influential movement of ideas in the opposite direction to that suggested above, from society to science. This directional pull has been highlighted by scholars who have been quick to use it as evidence of the inadequacy of the idea of objectivity within racial studies and science in general. Thus Kenan Malik has argued that the positions which were and are taken by scientists on the issue of race can only be understood from outside the scientific context. To Malik, there 'was no dynamic within scientific discourse … that lent itself to racial theories. Social and political factors determined the relationship.'[10] Similarly, Nancy Stepan has emphasised the influence of political ideologies over scientific affairs, noting, on the stance of scientists to the Eugenics movement in Britain, that 'science in itself did not determine the position biologists took'.[11]

Whilst it is pertinent to highlight the impact of politics on the science of race, it is equally important to recognise that science in turn has had an impact on politics, albeit in an inconsistent and indirect way.[12] Goldberg has argued in this context that 'race in its various articulations has served not only to rationalise already established social relations but to order them'.[13] Ultimately, Barkan suggests that it is impossible (or at least inadvisable) to try and work out which (of science and politics) has the guiding influence over the other. Instead, he argues, it is more pertinent just to realise that the two are 'strongly intertwined'.[14] Other scholars have shown a similar caution over trying to read direction into the relationship between science and society. For example, one influential study of science and its social context has suggested:

> Science does not stand outside of society dispensing its gifts of knowledge and wisdom; neither is it an autonomous enclave that is

now being crushed under the weight of narrowly commercial or political interests. On the contrary, science has always both shaped and been shaped by society in a process that is as complex as it is variegated; it is not static but dynamic.[15]

This study will explore in detail this confusing 'intertwined' relationship between science and society, whilst trying to avoid any kind of simplistic 'chicken and egg' analysis of whether one informed the other or vice versa. Instead, it will engage with the history of racial science in Britain, mindful of Roger Smith's criticism of the recent academic tendency to split history into 'two cultures', dividing scientists from the wider community.[16] Smith's challenge asserts that in fact a 'common context' was discernible between science and society in the interwar period and some years on either side of it, which united biological and social commentary in 'a shared and durable world of expression and judgement'.[17] McGucken had earlier argued that this was indeed the case, describing the involvement of scientists in social affairs as the 'principal characteristic' of what he labels as a 'social relations of science movement', between 1931 and 1947.[18]

This idea of a growing 'common context' of shared values between science and society has been hinted at in other recent scholarship, albeit focused on a later period. Nowotny, Scott and Gibbons have described the second half of the twentieth century as a time of social/scientific dialogue on a new scale, an age where 'society has begun to speak back to science'.[19] In earlier research, these scholars utilised the idea of 'mode 2' science to describe the changing relationship between science and society. A key aspect of 'mode 2' change is the primacy of social/scientific 'continuous negotiation' over the production of knowledge, which 'will not be produced unless and until the interests of the various actors are included'.[20] To explore this idea of negotiation, this book will use the history of British racial science between 1930–1962 as a case study to question the social impact of expert thinking on race as well as exploring in detail the political influences which may have impacted on racial science.

The 'post 1945' thesis

The choice of period under consideration here is based upon an awareness that the years before, during and after the Second World War witnessed an international crescendo of racial politics. The rise and fall of Nazism and the Holocaust, the struggle for an end to Southern segregation in the United States, the formal birth of the apartheid regime in

South Africa, the final collapse of the European colonial empires and the beginnings of mass Asian, African and Caribbean immigration to Britain, ensured a climate both in and outside science where the issue of race was a sustained matter of dispute and attention. This in itself is one reason to address this period, though it is not the only one. It is arguable that the horrific violence surrounding some of the events mentioned here (especially Nazi racial policy and the Holocaust) has done much to mask the realities of the changing dynamics of racial thinking in modern Britain. This study will delve into this period to question whether too hasty readings of the impact of international racial violence on British sensibilities have led to an erroneous historical understanding of the decline of race in British scientific and social discourse.

It seems possible that the vociferous denunciations of colonialism, segregation and especially Nazi racism before, during and after the war have served to mislead scholars into believing that scientific beliefs in race declined and disappeared during the interwar and war period. In his recent study, King has asserted that 'something like a consensus...exists on the demise of race as a valid scientific idea...between 1920 and 1945'.[21] There is certainly a case for King's contention as regards scholarship on British racial science. For example, whilst Barkan cautions against the simplistic analysis that 'once Nazi atrocities had been revealed, racism was rejected', he concurs that 'belief in the biological validity of race as a concept' was undermined 'during the interwar years'.[22] Indeed, the whole rationale behind Barkan's groundbreaking *Retreat of Scientific Racism* study holds that the decline of scientific racism can be charted between the years of 1918 and 1939. Nancy Stepan has claimed a similar role for this period, arguing that Nazi racism 'virtually destroyed' the credibility of the concept of race and asserting that racial science was a phenomenon which endured until 1950.[23] King himself also posits this kind of argument in his claim that by the end of the Second World War, 'a consensus firmly opposed to racial differences grounded in biology emerged among intellectual and academic elites'.[24]

In the years since Barkan and Stepan wrote their accounts of British racial science, other scholars have begun to question the extent of the 'retreat of scientific racism' in the war and post-war period. Marek Kohn's *The Race Gallery* and Bill Tucker's *The Funding of Scientific Racism* have both highlighted the continuation of racial science after the Second World War and argued that the idea of race remained a significant scientific and social issue.[25] The case made by these scholars is not that the war and the Holocaust did not damage race, only that it did not bring about the end of the concept in science. Paul Gilroy has suggested from a similar perspective that 'the special moral and political climate that arose in the aftermath of National Socialism and the deaths of millions was a transitory phenomenon'.[26]

In another challenge to the 'retreat of scientific racism' thesis, Kenan Malik has argued that the fate of race after the war differed in political and scientific arenas. Malik's case holds that whilst the Holocaust discredited the 'political use of racial science... its conceptual framework was never destroyed'.[27] By addressing the relationship between racial science and political policy in Britain, this book will consider Malik's contention as part of an attempt to move on historical understanding about what really did happen to racial science in this period.

The meanings of race

Sociologist Michael Banton's efforts to deconstruct the evolving meanings of race in changing scientific contexts have gone some way to providing a 'conceptual framework' in which this study of scientific racism can be set. Banton has argued that the study of race has evolved through three 'phases'. In the first phase, which he dates as lasting from the sixteenth to the eighteenth century, race was understood in terms of lineage, 'to designate a set of persons, animals or plants connected by common descent or origin'.[28] At this stage, Banton argues, 'race' was just one of a variety of words which could be interchanged with other terms like 'variety' or 'species'. In the second phase, the usage of the term changed, coming 'to signify a permanent category of humans of a kind equivalent to the species category'.[29] It was at this stage that the idea of race took on the meaning that was to have such influence in the twentieth century, the notion that differences between human groups were natural and permanent. Finally, this phase was followed by a third phase where racial ideas were superseded by new explanations of human difference and similarity which did 'not need any concept of race'.[30]

This book will offer an exploration of the transition between Banton's second and third phases. In particular, it will assess the idea, expressed by Banton and other scholars, that a key role in this transition was played by geneticists who, from the 1930s, 'provided explanations of the physical inheritance of characters' which superseded the typological racial explanations which had characterised second phase analysis.[31] It will argue that scientific challenges to race were not triggered in any direct sense by the ascendance of genetics or any other new scientific or methodological innovation but instead that politics played the key role in moving the science of race on from phase two to phase three. In making this case it will emphasise that the scientific abandonment of the idea of mental racial difference evolved at a different pace from the abandonment of the idea of physical racial difference.

By identifying a two-track development of scientific thought on physical and mental racial difference it is possible to gain a clearer picture of the shaping hand of politics and ideology in the evolution of post-racial science. Ideas about physical racial difference were not as politically inflammatory as were their mental equivalents. This is not to say that notions of physical difference did not lead to racial violence and discrimination. It is beyond doubt, for example, that fears about the potentially degenerative physical effects of sexual mixing with perceived lesser types shaped racial analysis across Europe and the United States as well as in Nazi Germany.[32] Similarly, notions of black (usually male) physical super masculinity played an important role in stigmatising black people and specifically in generating the allegations of sexual violence and misconduct which underpinned so much of the segregationist world view.[33] However, ideas about mental or psychological racial differences were powerful and potent on an altogether different scale in generating racial conflict and feeding racial politics.

It was the idea that other perceived races were inherently psychologically different that drove the racial violence which characterised this period, especially the Holocaust. Jews were racially demonised for their inherent perfidy, their craftiness, for their natural psychological tendencies towards clannishness and their lack of moral values. Africans and Asians were mostly characterised as primitive, barbaric, simple and uncivilised. Indeed every race was prescribed with its own mentality, which, like a physical difference, could pollute or undermine the body politic if allowed to enter the racial stream of the nation.

Because of the power of these kinds of ideas, scientific challenges made against race were often differently focused on physical and mental constructions. Typological inaccuracies about physical inheritance were challenged so that there was, by the 1960s, much greater recognition of the overlapping of physical racial potentials, but essentially the idea of racial physical difference remained a part of the tapestry of science throughout this period. In contrast, the idea of mental racial difference became increasingly taboo, though not (as we shall see) to the extent that mainstream biology was prepared to dismiss it entirely. Nonetheless, discrediting the idea of mental racial difference became the primary goal of anti-racial struggle. It is arguable in this context that the best way to gauge the post-war continuation of racial science and the extent of anti-racial change is by focusing on scientific challenges to this particular type of alleged racial difference. Thus, this study will concentrate mostly on changing ideas about mental racial difference in biology, anthropology and sociology before during and after the Second World War.

Deconstructing racial science and British society

The final section of this book will look in detail at the contribution made by social scientists to racial analysis in Britain after the war. In bulk, however, this study will focus its attention on natural scientists, especially biologists, and other scholars who addressed biological as well as social constructions of racial difference. In order to assess change and continuity in scientific attitudes, it will follow the histories of some of Britain's leading scholars of race in this period. Most of these men were biologists (notably Julian Huxley, Lancelot Hogben, J.B.S. Haldane and Reginald Gates), whilst others were anthropologists, sociologists and geographers. Focusing on leading protagonists will not entail the production of a history of great men of British racial science.[34] Instead, following the approach laid down by Gary Werskey in his *Visible College*, it will use leading players as a window into a wider history of social and scientific values.[35]

To achieve this goal, this study will also give attention to organisations (especially the Eugenics Society) which tried to influence the public and the government on racial policy. However, this book will not attempt to write another history of the Eugenics Society, partly because there is already a strong selection of books on this topic, but also in recognition of the fact that the bulk of the Society's concerns were not focused on racial matters in any obvious sense.[36] It will instead explore specifically the Eugenics Society's evolving attitudes towards race and immigration between 1930 and 1962, giving attention to Dan Stone's analysis that the Eugenics Society's lack of interest in race has been overstated by historians and challenging the idea that the racial views of members of the Society were, as has been erroneously argued by some, 'incidental to their eugenic concerns'.[37]

Understanding the Eugenics Society and its levels of social and political influence is no simple matter. For one thing, it is important not to read history backwards, and ascribe to the Society agendas and aims which are now assumed by a contemporary culture which does not perceive eugenics as legitimate either in terms of science or morality. Explicitly, the historian must be careful not to see eugenics, even in this late period, as the preserve of the political right.[38] Nor would it be accurate to cite statements by random Society members, or publications in the *Eugenics Review*, as necessarily being representative of the Society's agenda. Even by the standards of pressure groups and political parties, the Eugenics Society was born, and remained, a broad church. Within the Society, even issues related to core values (such as the promotion of birth control) were at times hotly disputed at the highest level.[39]

The term 'eugenicist' was in itself highly ambiguous and used to describe a myriad of conflicting agendas.[40] It can be said with confidence that the term was first employed in the 1883 writing of scientist Francis Galton and that the Eugenics Society (which first met as the Eugenics Education Society in 1907 under Galton's leadership) aimed to promote the importance of 'good breeding' as a vital national issue.[41] However, what needed to be done to secure 'good breeding' remained a long-running matter of dispute within the Society. For example, some British eugenicists, such as Caleb Saleeby, aimed to focus the new science on wider welfare provision to improve the nation's health. Saleeby's work called for the eradication of the 'racial poison' of alcohol and the establishment of free health centres.[42] Contrast Saleeby's views on social welfare with those of eugenics enthusiast, writer and politician Arnold White, and the enormous ambiguities and internal contradictions of British eugenics become clear. White argued that 'modern civilisation and philanthropy, on the whole, are hostile to conditions of sound national health' and that society must be 'content to see the idle perish'.[43]

Around these wide and often contradictory agendas, this book will argue that it is still possible to track governing ideas and concerns within the British Eugenics Society. It will seek out the dominant currents of racial thought in British eugenics, focusing in particular on the responses of the Society to contemporary political issues such as Nazism, US segregation and black immigration to Britain. As an overarching theme, it will focus on the Society's responses to the decline of race as a scientific and social idea in this period and will attempt to gauge its level of social and political influence.

This attempt to quantify and evaluate the level of influence of the Eugenics Society is part of a wider goal of this study, namely to improve historical understanding of the relationship between experts and the wider population as regards the issue of race. Jones has argued that the Eugenics Society can be addressed in this way as 'a nodal point at which political and social belief and the practice of science met'.[44] Smith's analysis of this social/scientific interaction (which has used the

Eugenics Society as a key example) highlights an idea that will run through this book, namely that this relationship should be understood less in terms of direct influence and more in terms of shared values:

> If historians agree that eugenics, understood as a specific social movement, did not have the influence in Britain once attributed to it, many related, if often rather vague, views linking biology and social policy were nevertheless held by people far beyond the direct reach of the Eugenics Society. Support for the notion that biology explains character, and that character, or the capacity for individuality, is both the key to social advance and an ethical ideal, was commonplace.[45]

Smith's idea of 'common context', already discussed here, denies science a lead role in opinion shaping and instead promotes the notion that scientific and social views are linked in a more complicated way. This book will explore this relationship and suggest that instead of leading society on racial ideology, science in fact operated in a shared discursive terrain; that the key scientific ideas that dominated expert writing on race in this period mirrored wider social racial thinking. The issue of immigration policy will be used in this study to reveal this relationship. Throughout the period under consideration, this book will consider the construction of immigration policy set against contemporary expert thinking on race, providing a window into the influence of racial science on policy and vice versa. To facilitate this goal, the final part of this introduction will outline the influence of racial ideas on early twentieth-century British immigration policy in order to provide a context for the period that followed.

The aim here is not to argue that popular and scientific constructions of race were identical. Political interest in the idea of race was not constructed exclusively or even predominantly on scientific foundations but was instead eclectically informed by a range of racialised ideas emanating from science, literature, religion and tradition, which Banton has described as 'folk concepts' of race. However, the line between science and society is, as we shall see, frequently blurred. For one thing, popular ideas or 'racial discourses' in society fed into science, influencing both questions and answers about race as a scientific concept, which in turn often fed back into popular discourse. Deconstructing the policies of immigration restriction from 1905 onwards seems indeed to reveal that the interest in race which had been mounting in British science since the latter parts of the nineteenth century mirrored growing social concerns about racial difference and the need for British racial preservation.

Race and immigration policy in early twentieth-century Britain

In 1905 the British government enacted the nation's first modern restrictive immigration law in the shape of the Aliens Act.[46] Retrospectively, and in the context of a century of restrictive legislation, the 1905 Act seems limited and moderate in scope. Despite its intention of controlling immigration to Britain, the Act only aimed to prevent the settlement of those immigrants deemed destitute; who would immediately become a charge on the state, as well as those who might perpetrate crime. For the most part, anti-immigration voices in Parliament focused their campaigning on these specific issues and not on immigration as a peril in itself. Even the vociferous pro-restriction MP William Evans-Gordon (who considered himself 'the father of the Aliens Act') explained the Bill as designed 'to rid our gaols of foreign criminals and our streets of foreign prostitutes and souteneurs'.[47]

But the 1905 Act (and the 1903 Royal Commission on Alien Immigration which preceded it) also reflected a growing social belief in the need to control more tightly the racial composition of the nation, a concern similar to that which led to the foundation of the Eugenics Society in 1907.[48] Worries about national degeneration and racial deterioration had become more prevalent in the years prior to the Aliens Act, especially as many national commentators explained away poor military results in the Boer War in the language of declining British racial stock.[49] In an era when Britain felt under increasing competition from European rivals, the need to protect the racial qualities which were perceived as having made Britain great in the first place became increasingly felt.[50] Accelerated Jewish immigration, fuelled by Russian anti-Semitism, exacerbated fears that British racial characteristics faced the threat of dilution and pollution by alien stock.[51] H. Lawson, the Liberal Unionist MP for Mile End, told the House of Commons that the arriving immigrants were 'the derelicts of Europe'. He cautioned: 'I think it is time we called halt to what is becoming a backward march to physical degeneration.'[52] The racial agenda of the 1905 legislation was indeed summed up by the Prime Minister himself at the end of the second reading of the Aliens Bill:

> If there were a substitution of Poles for Britons, for example, though the Briton of the future may have the same laws, the same institutions and constitution, and the same historical traditions learned in the elementary schools. Though all these things might be in the possession of the new nationality, that nationality would not be the same,

and would not be the nationality we should desire to be our heirs through the ages yet to come.[53]

The Aliens Act was extended in 1914, passing easily through Parliament only hours after Britain's declaration of war on Germany.[54] The new Act gave the government an awesome array of powers over 'aliens' in Britain, ostensibly to enable the control of war enemies.[55] The Home Secretary gained the right to place absolute restrictions on immigrant numbers, to control where immigrants did and did not live and demand that they register at frequent intervals, and to deport them from Britain if he saw fit. This law was designed to aid a state at war, but its powers were retained and cemented in post-war immigration legislation.

In fact the 1919 Aliens Restriction Amendment Act actually tightened the earlier law by further restricting alien rights to enter the UK and to own property, land and businesses.[56] The debate surrounding the legislation revealed a belief amongst many parliamentarians that allowing the entry of foreigners would damage the racial fibre of the nation.[57] One Conservative anti-immigrant MP, Stanton, argued that allowing any aliens to remain in Britain would be 'a stain upon our British stock', whilst Pemberton Billing told the House that Britain risked becoming 'an asylum for the ne'er-do-wells and parasites of the world'.[58] In a later debate, the comments of another Member reveal the presence of pseudo-scientific constructions of race within parliamentary argument:

> These four limbs of the race – Saxon, Norman, Dane and Celt – have given the nation the power that it is today by the mingling of their strength. I am content to maintain our stock as nearly as possible from these four races.[59]

The detail of the 1919 Act suggests that the framers of this legislation bought heavily into prevalent social racial discourses. The new law prohibited immigrants from changing their names whilst living in Britain and outlined provisions to deport any immigrant who was suspected of inciting industrial or political disorder.[60] Whilst these restrictions were most obviously framed in fears that followed the Bolshevik revolution in Russia and its aftermath, they combined with the already mentioned prohibition of immigrant ownership of British business and land to reflect growing concerns about Jewish conspiracy and manipulation.[61] The law seems specifically to have been designed in order to counter the craftiness and treachery of the Jew; to prevent his slipping unnoticed into British society by changing his foreign name and checking his natural desires to take over British business and industry. The characterisation of the Jew as a revolutionary Bolshevik may also explain the draconian provisions for the imprisonment of 'seditious' aliens for up to ten years.[62]

Similarly, the 1919 Act provides some insight into the influence of racialised perceptions of black people. Whilst Article 5 of the legislation prohibited the employment of foreign seamen at lower rates of pay than British sailors, the same article made a revealing racial exception. Lower rates could be paid to aliens who were chosen to work 'in any capacity or in any climate for which they are specially fitted' provided that this rate was the same as 'standard rates for British subjects of that race'.[63] Allowing ourselves the presumption that it was black sailors who were perceived as 'specially fitted' for certain climates, the implication is clear. One rate of pay existed for British and alien white workers, another for British and alien black workers as the idea of race transcended the idea of nationality in the political imagination.

Other measures taken by successive British governments in the inter-war period reveal a reluctance to accept the idea of British citizenship for black people. The Aliens Order of 1920 and the Special Restriction Order of 1925 (better known as the Coloured Alien Seamen Order) provided the police with the de facto authority to register black Britons as aliens.[64] It is evident that these powers were utilised to just this effect. The 1920 Act gave authorities the right to examine or detain anyone who was 'reasonably supposed to be an alien'.[65] The 1925 order cemented this right by forcing 'coloured alien seamen' to register with the police, who then had the responsibility of monitoring and supervising their whereabouts.[66] In principle, neither of these Acts should have affected the lives of British black seamen. However, within a racial classification of Britishness, it is clear that being black was in itself considered 'reasonable suspicion' of being an alien.[67] After the 1925 Order, hundreds of black Britons were cajoled into registering themselves as aliens, in what can be perceived as a covert attempt to disrupt the establishment of British black communities by the state and the police authorities.[68] The *Journal of the League of Coloured Peoples* recorded how this legislation was utilised against the British black community. As late as 1935, a report from Cardiff noted how black Britons were 'forced by fraud to register as aliens'.[69] A further article by the same author concluded:

> The interpretation placed upon these legislative measures is that they automatically made every coloured seaman in Cardiff an alien,

irrespective of the fact that he was born a British subject and had documentary evidence of his nationality.[70]

The proceedings of the 1921 Aliens and Nationality Committee seem to corroborate the case that the government perceived race as the primary designator of British nationality. The committee discussed the possibility of allowing access to the British mainland only to 'British born subjects of European race'.[71] This idea was rejected in the committee as too controversial, but it goes some way to show the will of some policy makers. Further evidence supports the contention that the Aliens and Nationality Committee placed the importance of race over that of citizenship. A memorandum from the Department of Overseas Trade to the committee argued that the treatment of foreigners in Britain should depend not on their nationality but on their racial origin:

> A very large proportion of naturalized British subjects described as of Austrian origin are in fact by race Czechs, Poles, Polish Jews or Slavs of some kind, and in such places there would appear to be no reason for regarding the persons in question differently from any naturalized British subject of friendly origin.[72]

Here race and not nationality was perceived as answering the question of a person's real abilities and loyalties, an idea which would find resonance in both science and society in the years that followed. In the wake of the 1919 Aliens Restriction Amendment Act, the *Jewish Chronicle* critiqued the government's response to minorities in Britain, complaining: 'The anti-alien craze will not be satisfied at national differentiation. It is really racial differences that it would mark, perpetuate and penalise.'[73] As the century continued, faced with new challenges and problems, race was an idea that would continue to play a major role in shaping the reception of immigrant and minority communities in Britain.

2
Rethinking Interwar Racial Reform: the 1930s

Historians of scientific racism in Britain have consistently described the 1930s as a period of change. From a variety of perspectives they have argued that this decade witnessed innovative scientific approaches to the study of race and a corresponding decline in traditional racial scholarship.[1] Stepan describes the period as a time of 'transition and ending', Mazumdar, as an age when 'a new alignment of method and ideology' conspired to undermine racial science.[2] This chapter will focus on these ideas of 'transition' and 'ending', questioning the validity of the idea of biological rethinking on race in the 1930s. It will consider exactly what changes occurred in interwar racial science and why they did. It will argue that the parameters of scientific rethinking in this period need to be clearly understood amid what has been a teleological tendency to over-egg the pudding, to read back into history changing values and methods that really were not dominant in scientific approaches to race until far later in the century.

In a succinct expression of the dominant historical perspective, King has noted the manner in which 'recent studies fix on the years between 1920 and 1945 as the period in which race, racial difference, and racial hierarchy were largely discredited among intellectual and scientific elites'.[3] King reminds us that key discussions on racial science amongst geneticists and anthropologists in this period most often concerned the increasingly disputed veracity of ideas of racial difference and hierarchy. However, it is important to clarify the scope of this scientific discussion. In the 1930s, the existence of physical and temperamental differences between different races was seldom a matter of dispute.[4] Instead scientific debates probed the root and meaning of racial differences, focusing on the following two key linking questions which set the tone of debate in the interwar period and beyond:

1. To what extent are differences between races due to inherent or cultural factors?
2. Is it necessary to monitor or control the extent to which races mix together?

These questions did not challenge the core principle that races (and racial differences) existed, only the origins and meanings of these differences. Comprehending these limits to discussion is absolutely essential to understanding scientific racial study and its impact in the interwar period, for ultimately this was not a period when race as an idea disappeared in any sense, only a time when race was reconsidered and rearticulated.

Four racial field studies

The parameters of racial thinking in this period can be illustrated by an analysis of two terrains of disputed racial field research, both of which led to academic publications between 1925 and 1940. One area of research focused on the assessment of Jewish immigrant children in London, as part of a wider discussion about the impact of Jewish immigration on British society; the other, on the effects of black settlement in Liverpool. Two significant reports were published on the subject of Jewish immigrant children. The first was based on a study by the mathematician and eugenicist Karl Pearson and his assistant Margaret Moul which was conducted before the First World War and finally disseminated in the *Annals of Eugenics* between 1925 and 1927.[5] The findings of this report provoked another similarly constructed study by psychologists Arthur Hughes and Mary Davies, published as a pamphlet and in the *British Journal of Psychology* in 1927.[6]

The second terrain of racial research explored the growth of 'coloured' and 'mixed race' communities in Liverpool in the interwar period. In 1930, Muriel Fletcher, a probation officer trained in Liverpool University's School of Social Science, published a report on the 'colour problem in Liverpool and other ports'.[7] This research was commissioned after a meeting at the School of Social Science to discuss 'half-caste' children, attended by the University Settlement, the police and representatives of the University itself, addressed by Rachel Fleming (assistant to the physical anthropologist and geographer H.J. Fleure at the University of

Aberystwyth).⁸ Fleming's authority on the subject stemmed from research that she had recently conducted with Fleure into 'Anglo-Chinese' children, at the bequest of the Eugenics Society.⁹ The Liverpool meeting led to the establishment of an Executive Committee to look into the issue of 'half-caste' children, which became the driving force behind the Liverpool Association for the Welfare of Half-Caste Children, which commissioned Fletcher's report.¹⁰ In 1940, David Caradog Jones, a keen eugenicist and social scientist from the University of Liverpool, responded to Fletcher's findings in one of a series of reports about specific immigrant groups and integration in Liverpool, focusing on the economic status of the 'coloured' community.¹¹

Moul and Pearson conducted their research in the form of a medical examination of a sample of Jewish immigrant children at the Jews' Free School in Spitalfields in the East End of London in 1910.¹² The aim of this research was explicitly social and political as Pearson noted in his initial 1911 notes. Not only would a better understanding of the lives of immigrant children help to 'ameliorate their conditions', but such an understanding would enable Britain to 'take every possible precaution against the physical, mental, and moral degeneration of the race'.¹³ In the published version of Moul and Pearson's report, this desire to safeguard the racial stock of the nation was highlighted even more explicitly as the primary agenda of the research:

> What purpose would there be in endeavouring to legislate for a superior breed of men, if at any moment it could be swamped by the influx of immigrants of an inferior race, hastening to profit by the higher civilisation of an improved humanity.¹⁴

To Moul and Pearson, the intellectual political rationale of their research was clear. If the immigrant Jewish children were shown to be superior to their non-Jewish equivalents (and capable of assimilating with them) then their presence in Britain should be encouraged and further immigration solicited. If, on the other hand, the study showed them to be inferior or incapable of assimilation then steps needed to be taken to restrict or curtail Jewish immigration and preserve British racial stock.¹⁵ The research was part funded by Jewish magnates who must have been similarly aware of these potential political implications and perhaps hoped that the authors would, in their analysis, help to justify the place of Jewry within the racial ranks of the British nation.¹⁶

It is important to understand these goals within their political context. Moul and Pearson's initial field research was carried out in the wake of a decade of frequently frantic British debates about the implications of immigration. These debates partially culminated in Britain's first Aliens Act in 1905, legislation that was designed to curtail, albeit implicitly, the immigration of poor and criminal Jewish migrants. The climate of publication in 1925 was similarly politically charged. Questions about the desirability of Jewish immigration from Eastern Europe were still very much on the political agenda.[17] Moul and Pearson stepped into this atmosphere with the confidence that scientific method could determine immigrant desirability.

The conclusions of the report were quite clear. Moul and Pearson considered that the surveyed Jewish immigrant children 'were inferior in the great bulk of the categories dealt with' encompassing both physical and mental tests.[18] The scholars' final conclusions were equally certain: 'Taken on the average, and regarding both sexes, this alien Jewish population is somewhat inferior physically and mentally to the native population.'[19] Given these findings, Moul and Pearson did not hesitate to assert what were for them the obvious political implications:

> The welfare of our own country is bound up with the maintenance and improvement of its stock, and our researches do not indicate that this will follow the unrestricted admission of either Jewish or any other type of immigrant.[20]

This conclusion was underpinned by the belief that the observed differences between Jewish and non-Jewish children were the product of heredity and not environment. If Jewish underachievement could be explained away as rooted in social disadvantage then there would have been no validity to a case to restrict immigration and a different agenda, to improve living standards amongst immigrants, would have been the logical outcome. Moul and Pearson were however adamant that no such social amelioration would improve the immigrant Jewish stock in question:

> Our material provides no evidence that a lessening of the aliens' poverty, an improvement of their food, or an advance in their cleanliness will substantially alter their average grade of intelligence, and with it their outlook on life...The native level is not a product of atmosphere, but of centuries of racial history, selection, hybridisation and extermination.[21]

This report caused considerable consternation within British scientific communities because of its conclusions about the racial worth of the

Jewish immigrants in question.[22] Within two years, Davies and Hughes had published a rival report addressing a similar comparison of the intelligence of Jewish and non-Jewish schoolchildren. It is revealing of the perceived political ramifications of this scientific discussion that this research was funded indirectly by the Board of Deputies of British Jews through the Jewish Health Organisation of Great Britain.[23]

The Davies and Hughes report firmly rebutted Moul and Pearson's findings, arguing that Jewish children were in fact more intelligent than their non-Jewish peers. Conducting their research through a series of tests in several different schools the authors concluded: 'The most striking feature is the general superiority of the Jewish children, whether boys or girls, alike in intelligence and in attainments in English and Arithmetic, at almost every age from 8–13, and in every type of school.'[24] This report made implicit criticisms of Moul and Pearson's methodology, noting that the earlier study had conducted research only in the Jews' Free School and had relied solely on simple observations. Davies and Hughes argued: 'The [Jewish] superiority in intelligence revealed by the tests is confirmed by the independent estimates of the teachers...In no other school do we find Jewish children as poor in intelligence as Pearson and Moul found in the particular school studied by them.'[25]

Comparing these two reports tells us much about the state of racial science in Britain prior to the accession of Hitler. There is a case for arguing that the Davies and Hughes report offered a firm refutation of Moul and Pearson's racial analysis. Describing the report, Richards has thus argued: 'If classifiable on methodological grounds as Race Psychology, this paper was thus hardly promoting anti-Semitic racism, on the contrary it was Jewish-funded and its findings flattering.'[26] Furthermore, Davies and Hughes ascribed a greater significance to environmental factors in their report than Moul and Pearson had done. As such, their report could be read as a critique of theories of inherent racial difference. Davies and Hughes argued that a key disparity between their research and Moul and Pearson's was their predecessors' focus on newly arrived immigrants, who were disadvantaged in terms of environmental factors (like poverty and language proficiency):

> ...most of the children examined in the Pearson-Moul investigation were children of recent immigrants and 28.3% of them had been

born abroad. Of the Jewish children tested in the present investigation only 3.1% were foreign born.[27]

However, the methodological queries posed in the Davies and Hughes report did not amount to a significant challenge to the idea of inherent racial difference. Whilst Davies and Hughes did want environmental issues factored into the immigrant assessment equation, they shared much more in common with Moul and Pearson's ideology and methodology than may seem at first glance to be the case. For one thing, the Davies and Hughes research shared Moul and Pearson's assumption that there was methodological validity in a racial assessment of an immigrant group set against the host community and that there were likely to be quantifiable differences between the two races. It is significant that the Jewish Health Organisation of Great Britain commissioned Davies and Hughes on the advice of Cyril Burt, who then assisted the authors with their report.[28] Burt was the official psychologist of London County Council and was as such responsible for intelligence tests across London schools.[29] He was renowned as a determined believer in the hereditary inheritance of intelligence, and has been described by Stepan as the man who 'effectively ensured that the Galtonian, psychometric tradition of the study of mental differences became the dominant traditions of British psychology'.[30] The approach to Burt thus indicates that there was no inclination on the part of Davies and Hughes's paymasters to dethrone the idea of mental racial difference through the new report, only to secure different results.

The shared terrain of belief in inherent racial mental differences between Moul and Pearson and Davies and Hughes was made very clear in another paper addressing the differences between 'Jews and Gentiles' published by Arthur Hughes in *Eugenics Review* in 1928. In this paper Hughes argued:

> Even the most ardent internationalist would admit that if races ought to be reckoned as equal, it is on a basis of diversity of usefulness and not of identity of capacity...That Jews differ from non-Jews is generally accepted as an axiomatic truth.[31]

To Hughes, whilst environment did play a role in shaping personal and group achievement, differences between Jews and non-Jews were inherent, and due to breeding. He cited 'the unnatural selective influence of years of persecution' and the 'tradition for rich Jewish fathers to seek out men rich in learning rather than in goods' as reasons for Jewish racial superiority.³² His final conclusion held that environmental handicaps could only dull racial differences, not undermine them:

> The poorest Jewish children, according to our results, are so far superior to the poorest non-Jewish children as to be practically equal to the general non-Jewish average, a fact which points very definitely to 'racial' superiority.³³

Ultimately, these two rival reports opposed each other in terms of conclusion and implication, not in terms of core methodology or differing beliefs in the importance of race. Davies and Hughes felt Jewish immigration would enhance the nation, Moul and Pearson that it could damage it. Neither report challenged the idea that Jews were racially different from non-Jews and neither disputed the idea that immigration to Britain could and should be assessed in terms of racial impact and eugenic desirability.

Similar to these studies of Jewish immigrants in London, the Fletcher and Caradog Jones reports in Liverpool were focused on the effects of integration, specifically on the results of a generation of urban cohabitation between white Britons and black immigrants. Like the field research on Jews, the Liverpool projects had a clear political agenda and context. In 1919 Britain had erupted in a series of race riots of varying severity.³⁴ Amongst the most serious took place in Liverpool, where rioting caused the death of a 24-year-old Bermudan named Charles Wooton.³⁵ Instead of being perceived as victims of racial assault, blame in the wake of the rioting was largely put upon the nascent black community, who suffered the brunt of arrests and punishment along with increasing calls for their repatriation.³⁶

The events of 1919 fed into and fuelled existing debates about the desirability of there being a settled black population in Britain. This growing discussion is reflected in tentative government steps in the 1920s to restrict black settlement (outlined in the introduction) and in the actions and agenda of various political/community groups. Organisations like John Harris's Anti-Slavery Society focused their efforts on keeping black and white communities separate and on 'helping' black Britons to return to their countries of origin.³⁷ It was seemingly with a similar goal in mind that the Liverpool Association for the Welfare of Half-Caste Children commissioned the report that became known as the Fletcher Report in 1927.³⁸

This Association had decided much about the 'half-caste' children that it set out to investigate before Fletcher was even commissioned.

When Rachel Fleming first addressed the Association in 1927 she told them unequivocally that the inter-breeding of white Britons with black seamen was creating mentally and physically disadvantaged children, 'born with a definitely bad heredity'.[39] According to Fleming (who had, with Fleure, recently conducted research into Anglo-Chinese children), the offspring of black–white unions inherited 'disharmonious mental and physical traits', creating a problem which 'depresses very considerably the life of the Docklands population of Liverpool'.[40]

Given this tone of inception it is unsurprising that the Fletcher Report reads as a polemical assault on the very existence of black communities in Britain, in particular decrying the development of a 'half-caste' population in Liverpool. The tone of the report was set in its foreword by P.M. Roxby, professor of geography at the University of Liverpool and chairman of the Executive Committee set up to investigate the issue of 'half-caste' children. Roxby at the outset of the report described the existence of such a population as 'a real social menace', arguing: 'it would be a sin against posterity to allow a proved evil of this kind to remain unchecked'.[41]

Inside the report itself Fletcher made reference to the previous attempts to restrict black entry to Britain. She described the 1925 Seamen's Order as 'useful', only lamenting that 'in Liverpool it has had but little effect as it applies only to aliens, while the majority of West Indian firemen on the ships of Liverpool firms are in possession of British passports'.[42] Her ultimate conclusion held that it was necessary to exclude black labour, specifically 'to replace all coloured firemen by white on all British ships coming to this country'.[43] Fletcher's dogmatic opposition to a continuing black presence in Liverpool was rooted in her hostility towards interracial sexual contact and her beliefs about the inferiority and inherent flaws of 'half-caste' children. As we will see later in this chapter, scientific debates about race in the 1930s focused more on this issue than any other.

Fletcher's investigation emphasised the sexual differences of black men whom she presented as promiscuous and deviant. She argued that:

> In Liverpool there is evidence to show that the negro tends to be promiscuous in his relations with white women. The women say that they are approached by other men if their husband's [sic] are away, in many cases the man who comes being the husband's best friend.[44]

Fletcher argued that not only was the black man promiscuous, but he lacked all notions of sexual decency, even trying to pay women to 'consort with their daughters'.[45] The report produced an image of a super-sexual black masculinity with passions that could not be sated in British society: '...their [black men's] sexual demands impose a continual strain on white women'.[46]

This analysis led to the corresponding assertion that no 'normal' white woman would partner a black man. Once a woman did so, she effectively abandoned ordinary sexual practice and became a part of the debauched black sexual world. 'Once a woman has lived with a coloured man, the house appears to become a sort of club for any coloured men in port. A white woman who has once mixed with coloured men is unable to break away and is never safe in a house if her husband is away for any length of time.'[47] It was seemingly inconceivable to Fletcher that an interracial relationship could be healthy or moral. Instead she perceived these unions as rooted in sexual depravity.[48]

Fletcher saw the children of mixed-race relationships as degenerate in line, as we shall see, with much of mainstream scientific racial opinion. 'The children seemed to have frequent colds, many were also ricketty, and several cases were reported in which there was a bad family history for tuberculosis.'[49] Not only did the unions produce physical disharmonies, but the report alleged that the children were also mentally inferior and infected with their parents lack of sexual morality.[50] Ultimately, Fletcher's report left little to the imagination concerning her views about the racial desirability of a black community in Liverpool. To this author and her sponsors, black Britons were a racial menace and their removal was crucial if the stock of the nation was to be preserved.

Caradog Jones's report, ten years after Fletcher's, was born into a not dissimilar atmosphere. The report was commissioned by the Liverpool Association for the Welfare of Coloured People, an organisation that shared both many of the goals and the personnel of the Association for the Welfare of Half-Caste Children. The foreword to Jones's report was written, like Fletcher's, by Percy Roxby from the University geography department.[51] However, Caradog Jones's report did offer something of a quiet critique of its predecessor. For one thing, Jones's project employed a black research assistant, thus reducing (as Roxby noted in his foreword) the 'fear of misunderstanding' that could occur when the investigator was white.[52] Secondly, the new report was firmly focused on the employment situation of 'coloured' people, enabling a different tone to Fletcher's emotive analysis of 'miscegenation'. Whilst Jones did not criticise explicitly Fletcher's report, his comments on the 'previous investigation' do show implicit discord as well as some significant differences in view. Jones noted that despite overlapping membership, the

Board behind his report 'must not be held responsible for the conclusions reached by its predecessor'.[53] Furthermore, Jones argued the case for his study by asserting a quiet lack of confidence in Fletcher's findings. In Jones's words, whilst Fletcher's conclusions may have been 'true then, they may not still be true'.[54] A subtle difference in outlook concerning the idea of black people in Britain is also detectable between the two investigators. It is doubtful that Fletcher could have stretched herself to share Jones's premise: 'It must not be presumed that "coloured families" ... are necessarily unwelcome in this country.'[55]

Despite these differences, Caradog Jones's report did not firmly challenge Fletcher's image of Liverpool's black population. Like the earlier report, the starting point of Jones's study was again the premise that black Liverpudlians were essentially alien, and needed to be assessed and monitored as a distinct social problem.[56] Furthermore, Jones seems to have shared Fletcher's belief in the physical and mental weaknesses of mixed race children. The final section of his report revisited Fletcher's conclusions on the 'Problem of the Half-Caste Adolescent'.[57] At the end of this section (which is also the end of the report) Jones leaves the reader with what he describes as 'illustrative extracts' about some of the assessed mixed race children, entailing six descriptions of 'half-caste' adolescents. In these descriptions, Jones presents an image of the 'half-caste' child as mentally and physically inadequate. Two of the children are recorded as suffering from mental disabilities, both described by Jones as 'rather backward'. Another three of the six are recorded as physically unfit for work, with conditions ranging from bad nerves, rheumatism and weak chests to eye defects. Only one of the group, described by Jones as 'not very communicative', was not dismissed as either physically or mentally unwell.[58] In light of this conclusion, it seems that generally Caradog Jones shared Fletcher's opinion, if not about the desirability of a black population in Britain, then about the effects of racially mixing with this population.

The importance of these four reports to this study is not that they represent all the racial opinions of biologists or social scientists in the 1930s but that they highlight something of the parameters of the race debates that were taking place across British scientific communities. Most of all, Fletcher's tendency, noted by Rich, to portray 'social and economic conditions... within hereditarian and eugenical terms', reflected a wider trend across British racial research.[59] Biologists would spend much of the next ten years arguing about the social effects of racial mixing and the economic and political implications of multiracial societies. Whilst Caradog Jones and Davies and Hughes may not have shared Moul and

Pearson's and Fletcher's views about the groups they were investigating, it is clear nonetheless that the idea that Jewish and black immigrants and minorities represented different and alien racial communities was to varying extents shared by all of these researchers.

As was suggested at the beginning of this chapter (and as has been shown here in the two sets of reports) racial research in the 1930s began to question the origins and importance of racial difference but rarely disputed the validity of the idea itself. If scientists in the 1930s increasingly had doubts about what race meant, there was near total consensus that it continued to mean something. Nonetheless this uncertainty about 'origins and importance' (evident here in the limited challenges posed by the Caradog Jones and Davies and Hughes reports) did signal something of a sea change. This study has argued so far that challenges to scientific racism in this period must not be exaggerated. It will now explore the changes that did occur in pre-war biological thinking on race.

Changing methods and changing politics

The issue of why scientists began to rethink racial issues in the 1930s has been a matter of historical controversy. A series of scholars have highlighted a gradual period of change in scientific thinking on race throughout the interwar period. In this school of thought the increasing influence of left-wing ideologies amongst many of Britain's leading scientists is emphasised as the key driving force behind changes to scientific racial analysis.[60] Socialist ideas, it is argued, sat uncomfortably with racial science in this period, which served through eugenical theory to justify policies which were mostly anathema to socialism, such as class hierarchy and colonialism. Mazumdar amongst others has thus contended that it was issues of class (more specifically, a fear of the working classes) that drove an essentially right-wing interwar British eugenic agenda. Separating the programme of British eugenicists from European colleagues, Mazumdar has argued:

> In each country, the eugenicists' *Wunschbild*, their ideal type, and its negative image, were determined by national background and historical context. In Britain, it was the casual labourers or pauper class whose low intelligence and high fertility were dangerous to society, as it had been throughout the nineteenth century.[61]

If this was indeed the case then it is easy to understand how left-wing scientists became alienated by eugenics and perhaps challenged thinking on race as a result.⁶² From a similar perspective, Stepan has shown that the 1920s saw the growth in Britain of a school of left-wing scientists whose opposition to scientific racial thinking was mostly rooted in leftist hostility to class prejudice, the British Empire and colonialism in general.⁶³ There is plenty of evidence that leftist political ideology did have a strong impact on the racial outlook of leading British natural scientists. J.B.S. Haldane, for example, voiced the idea that state intervention on racial policy was unscientific and actually rooted in a material desire to entrench inequality:

> The Germans have a right to rule others because they are a superior race, and the Jews must be expelled because they are inferior. The same sort of theories are used by the British in India, and by many of the whites in South Africa and the Southern States of the USA.⁶⁴

Similar views were expressed by biologist Lancelot Hogben who also saw the race issue both in Nazi Germany and in the British Empire as a matter of class struggle. Hogben argued that there was a 'parochial distinction between *Rassenhygiene* and its sister cult in Britain. In Germany the Jew is the scapegoat. In Britain the entire working class is the menace.'⁶⁵

Challenging accounts of racial change that prioritise the role of the leftist class agenda, Stone has highlighted the similarities between British and continental eugenic movements and has reminded researchers that thinking on racial difference was in fact as 'integral to the world view of the British eugenicists' as concerns class.⁶⁶ For scholars such as Stone, scientific rethinking on race in the interwar and wartime period can only be understood in the context of Nazi racism and the Holocaust. This understanding of racial change in the interwar period is the most prominent amongst scholars of race and science. As has been noted in the introduction, Barkan begins his book on scientific racism by highlighting what he perceives as the dominant narrative that 'Once Nazi atrocities had been revealed, racism was rejected.'⁶⁷ A recent biographical account of J.B.S. Haldane has highlighted this view. Having noted that J.B.S. wrote some highly racialised scientific analysis in the 1920s, we are then told: 'Later on, in the light of Nazism, he began to write against racism.'⁶⁸ Describing the changing racial perspective of Julian Huxley, Werskey has similarly noted the centrality of the Third Reich in shaping scholarly views before the war: 'Huxley's views on eugenics underwent a...transformation, in part because Lancelot Hogben did not hesitate to inform his friend that, in propagating them, he was acting like a Neo Nazi.'⁶⁹

These two schools of thought are not mutually exclusive. It is unarguable that J.B.S. Haldane, for example, reformed his writing on race as a result of both his growing belief in Marxism and his abhorrence of Nazism. Highlighting the reforming power of both socialism and anti-Nazism Stepan has thus argued: 'At the very least the debate about the concept of race inside and outside the scientific community left a question mark over its scientific credentials, even before the Nazi eugenics programme and Nazi anti-Semitism virtually destroyed its credibility.'[70] In the light of ongoing historical uncertainty a closer examination of British biology in the 1930s is necessary to show how and why scientific thinking on race changed in this period and allow a better understanding of the interplay of class and racial oppression as factors in changing scholarly views. It is to this detailed examination of biology that this chapter now turns.

Before considering the impact of these political matters on science it is perhaps pertinent to highlight that some of the reasons why scientists rethought racial ideas in this period were not ideological but rooted in academic innovation, important changes in methodological approach which made an impact on the way that scientists approached race as a topic. The kinds of approaches that had dominated racial study in the mid to late nineteenth century came under attack in the interwar years in a way that would substantially change the practice of science for good.

The patriarchs of racial science in the earlier period had been physical anthropologists like John Beddoe and Robert Knox; scholars who, despite holding radically different views about race, both believed that observed physical and temperamental differences could form the basis of a racial investigation.[71] The rise in the early twentieth century of biometrical research (a field led from University College London by Karl Pearson) put pressure on the analyses of these nineteenth-century anthropologists.[72] Pearson believed that the idea that simple observation of individuals could form the basis of any racial pattern was deeply flawed. Instead, he argued that racial trends were only observable through the analysis of group statistics and thus that the new science of eugenics needed to be rooted in statistical study, not crude observation. Pearson wrote in a 1909 paper: 'This transition from declamatory assertion to statistical proof is the characteristic feature of eugenics.'[73] Pearson did not intend for this new biometrical approach to undermine racial research.[74] On the contrary, he believed that the improved method was necessary in order to facilitate innovative and more accurate approaches to the study of race. In his words: '...great statistical advances had to be made, before it became possible in an effective and not merely periphrastic manner to study those agencies which may improve or impair the racial qualities of future generations'.[75] Nonetheless, by challenging crude, observation-based physical racial anthropology, Pearson unwittingly began to unravel the idea of racial difference.[76]

The rise of biometrics was rooted in scientific attempts to improve the potential applicability of Darwin's theory of evolution by analysing statistically observable continuous variations. This method was vigorously challenged by another school of British scholars made up mainly of zoologists and botanists. Following the rediscovered theories of Gregor Mendel, these scientists explained heredity and inheritance in terms of discontinuous characters, as Mendel had famously done with his peas.[77] The Mendelians believed in research through thorough experimentation, keen to avoid what Lancelot Hogben would later describe as the 'algebraical weeds' of Pearson's biometrical approach.[78] In order to promote rigorous experimental research, especially in zoology and botany, a group of like-minded British scholars founded the *Journal of Experimental Biology* in 1923. During the Christmas vacation of that year, the Society for Experimental Biology staged its inaugural conference at Birkbeck College. The 'Founding Fathers' of the Society were to become some of Britain's foremost biologists in the interwar and post-war period; they were Julian Huxley, J.B.S. Haldane, Lancelot Hogben and F.A.E. Crew.[79] The emphasis they placed on experimental rigour made an impact on the way that scientists approached racial study. For a start, as with biometrics, it led to criticism of earlier (now perceived as methodologically unsound) attempts at racial classification. As Hogben argued in a 1931 essay:

> ...the races into which mankind is divided by the ethnologist are based upon morphological distinctions. Little effort has yet been made to envisage their significance in the light of what has been learned by the pursuit of experimental methods.[80]

Hogben concluded that 'experiment and experiment alone' was needed to generate answers about the real nature of racial difference.[81]

Neither the biometrical nor the Mendelian schools were overly inclined to challenge the racial prejudices of science in the interwar years. It was unintended if nonetheless important that methodological conflicts between the two ensured a healthy level of academic rigour and critique which made it harder for all scientists to make racial claims that they could not substantiate. Stepan has thus argued:

> ...the biometrical-Mendelian dispute in Britain probably helped to quicken the pace of scientific criticisms of eugenics... For if the biometricians and the Mendelians could both claim to have a true science of heredity, were not the very scientific credentials of eugenics thrown in doubt?[82]

Without any deliberate inclination to undermine race as a concept, scientists in the interwar period added a new level of sophistication to racial research which would in time call into question scientific racial beliefs. Amid calls for both statistical and experimental accuracy, theories of racial difference were much harder to prove, especially in the acrimonious atmosphere of academic rivalry.

Although methodological changes created an atmosphere which facilitated reform it did not drive it forward, for whilst some Mendelians and biometricians utilised the new approaches to challenge scientific racial thinking, others did no such thing. Ultimately, scientific advances allowed for new emancipatory approaches to race, but also left staunch conservative thinkers plenty of room to manoeuvre. Barkan has concluded: 'Population genetics and the new evolutionary synthesis could lend credibility to either a racist or an egalitarian interpretation.'[83] Science left scientists with the scope to pursue various interpretations within new modes of thinking, enabling political beliefs to be all-important in setting the direction and tone of 1930s race analysis. The rise of Nazism, its racial abuse of minorities and the threat of war created a new climate of debate in British racial scholarship, not by leading to any scientific breakthrough or methodological change but by bringing the political perspectives of scientists to the fore.

Perhaps the first example of the changes in British race science that were generated by Nazism can be seen in the 'Race and Culture' conference of 1934. This conference was hosted by the Royal Anthropological Institute and the Institute of Sociology to discuss the issue of racial difference and was attended by British scientists from across the political spectrum, including conservative eugenic thinkers such as Reginald Gates and progressive left-wing theorists like Haldane and anthropologist Raymond Firth from the London School of Economics. The gathering achieved little in the way of consensus about the importance of race in science. Nonetheless, it served as an important symbol of a British scientific reaction, an intellectual response to the politicisation of racial issues in Nazi Germany. The conference publication recorded that the meeting had been necessary, as 'recent events have awakened the interest of large numbers of people in such questions'.[84] The merging of sociological and anthropological scientific communities in the conference in itself indicated a desire to develop a British united front as a challenge to the dogmatic racism of the Third Reich. Endelman has noted that the anti-Nazi anthropologist Charles Seligman 'successfully lobbied' the Royal Anthropological Institute into hosting the conference with this goal in mind.[85]

Many of the attending academics clearly wished to use the conference (as Seligman intended) to voice their opposition to Nazi racial thinking by promoting more critical and subtle scientific engagement with the concept of race. This progressive scientific perspective challenged the idea that race was a usable or stable concept and prioritised environmental explanations for cultural diversity between European populations. In doing so it was somewhat reminiscent of the kind of challenge to traditional racial analysis posed by Davies and Hughes (and later Rumyaneck) to Moul and Pearson's research a few years earlier.

Anatomist and anthropologist Grafton Elliot Smith argued at the conference: 'the acquisition of culture is not due so much to innate qualities as to historical circumstances and quite arbitrary factors'.[86] Raymond Firth endorsed this idea, noting that 'inter-racial aversion is a cultural product, not an innate disposition'.[87] J.B.S. Haldane contended that whilst some human characteristics were to his mind 'innate', race was not a usable concept when considering European populations:[88]

> ...this kind of classification will not work for the sub-division of such an area as Europe, and there is no reason to suppose that it would have worked at any time in the past.[89]

It is important not to exaggerate the extent to which these opinions formed a challenge to traditional racial study. Elliot Smith's above pronouncement that the acquisition of culture was 'not due *so much* to innate qualities as to historical circumstances' (my italics) is notable for its qualification and illustrative of the progressive challenge at the conference. Statements such as these did not attack the idea of inherent racial difference as such, but merely argued that race in itself was not the primary factor in shaping the cultural behaviour of a population. Haldane's intervention must be read in the same way. His comment (that Europe could not be divided, as the Nazis would have wished, into distinct races) did not signal that Haldane did not believe in racial differences.[90] It only indicated that he did not generally accept the idea of pure racial types in Europe. In the context of these limited criticisms of the use of race in science, Barkan's division of conference attendees into 'racists and non-racists' seems too simplistic.[91]

The 'Race and Culture' conference was unable to agree on a new progressive racial analysis amid the presence of two highly conservative racial thinkers, Reginald Gates and the anthropologist George Pitt-Rivers. Their presence at the conference highlights deep political divisions within British scientific communities and encapsulates the confusion and disagreement which dominated British scientific thinking on race in the interwar period. Gates and Pitt-Rivers did not share the inclinations of some of the other scientists at the conference to try to undermine the racial policies of Nazi Germany. As we shall see later in this study, Gates blamed Jews for their persecution under Nazism, before, during and after the war.[92] Pitt-Rivers was an overt Nazi sympathiser who was eventually interned in 1940.[93]

With no inclination to support the progressive views of Haldane, Firth and Elliot Smith, Pitt-Rivers and Gates used the conference to pursue more traditional racial arguments. These arguments mostly focused on the idea of physical (not psychological) racial differences between populations. However, scratching the surface of this perspective (and placing Gates's and Pitt-Rivers's comments in the context of their wider scholarship) indicates that conference statements on physical type were only the outer shell of deeply entrenched racist and anti-Semitic views.[94] As Barkan has concluded, 'these scientific theories accentuated the racialist interpretations and hence were compatible with Gates's and Pitt-Rivers's political alliances'.[95]

Pitt-Rivers directly challenged Elliot Smith's contention that cultural behaviour was not rooted in 'innate qualities' and asserted that whilst it was incorrect to argue that 'Aryan' was a real racial category, the terms 'Nordic, Alpine, Mediterranean' did describe 'recognisable and measurable...existing race-types'.[96] Gates used the forum to pursue the antiquated racial argument that different populations around the world were actually different species of animal.[97]

> If the same criteria of species were applied to mankind as to other mammals, it appears that the White, Black and Yellow types of man at least would be regarded as belonging to separate species.[98]

Whilst Grafton Elliot Smith reasoned that it was 'important not to fall into the common error of confusing race with nationality', Gates countered with the seemingly opposite idea, that 'the migrations and mingling of types has produced many new races, a process which is now going on more rapidly than ever before'.[99] Pitt-Rivers reinforced this theory by arguing that German attempts at Aryanisation could form a 'crucible' out of which 'new races or race-types evolve'.[100]

Amid political discord the 'Race and Culture' conference failed to produce any kind of definitive statement which could set out the stall of British scholars on the issue of race. It did, however, reveal a newly bolstered interest in racial studies in Britain, and a greatly increased commitment (amongst some British scientists) to take on the racial dogmatism of Nazi Germany. The publication of Julian Huxley and A.C. Haddon's *We Europeans* in 1935 provides a much clearer indication of this new commitment.

We Europeans set out to undermine the Nazi regime's use and understanding of the concept of race in European populations. Huxley recalled the project in these terms in his autobiography:

> It gave me a particular satisfaction to put this scientific spoke into Hitler's wheel, and to do something to stop his irrational anti-Semitism from spreading into Britain under Oswald Mosley's influence.[101]

Huxley and Haddon, as co-authors, were ideally suited to such a task. Straddling the fields of anthropology and biology between them, both well respected as moderate liberal-leaning men, they provided a dream team of British scientific impartiality. However, as various historians have noted, the actual group involved in the writing of *We Europeans* was much larger.[102]

Looking at the correspondence surrounding *We Europeans* it becomes clear that there were several authors. Letters to Julian Huxley from his assistant Phyllis Coomb at King's College London, reveal as much. Coomb wrote to Huxley three times in January 1935, trying to iron out confusion about who was supposed to be writing the various sections of *We Europeans* and how the book was coming together. Her letters suggest that the key decisions about the book were made by Huxley, Charles Singer and Charles Seligman. The letter below shows that these three retained an editorial role in the project, not just writing various sections themselves but correcting and commenting on those of others:

> I have now finished the corrections of the Race book, and have sent one copy round to Pinker, together with your remarks about the

articles from separate chapters. The other copy I am keeping here for the present. By the way, what has happened about Chapter 8 (Europe overseas – according to the list of chapters)? There doesn't seem to be one. Is it a chapter that is being done by somebody else, and what am I to do with it when it appears, or have you got it, or what? I have written to Singer and Seligman asking for any help about this, because I can so far find no reference at all to it.[103]

In other correspondence between Coomb and Huxley in January 1935 the important role being played by Singer and Seligman in the development of the text is highlighted further. 'The copy of the Race book for Seligman did get done, and he and Singer are going to see that all further corrections are made quite distinct from any others.'[104] In another letter sent the following week, Coomb highlighted the role of eugenicist and anthropologist Alexander Carr Saunders, who was fully credited as co-writer in later editions, as the author of the missing chapter on 'Europe overseas'.

I had a card from Seligman to say he had a draft of Carr Saunders' chapter, which he was taking to Cornwall to discuss with Singer this week; so that is getting on.[105]

From these letters it is clear that *We Europeans* had at least five contributing authors: Huxley, Haddon, Seligman, Singer and Carr Saunders. Certainly, it was not predominately the work of Huxley, even less so Haddon.[106] Why then was the work presented as co-authored by them? Barkan's explanation highlights the prevalence of fears about Jewish bias, similar to those which kept Seligman and Morris Ginsberg out of the 'Race and Culture' conference in 1934. Commenting on Seligman's contribution to *We Europeans,* Barkan has argued: '...Seligman's Jewishness supposedly compromised his ability to write a scientific monograph on race'.[107] As we shall see below, Haddon was deeply concerned about bias in the text and it seems likely that he may have insisted that only Huxley and his name were given as authors to prevent allegations of Jewish control. The addition of the name of the non-Jewish Carr Saunders, first as a contributor and then as co-author, seems to further suggest that it was Jewishness that was specifically a problem. However, a deeper understanding of Julian Huxley as a scholar suggests another additional explanation.

In 1941, as we shall see later in this book, Charles Singer wrote the Macmillan pamphlet on German racial science, *Argument of Blood,* under

Huxley's name. This happened because Huxley was asked to write the work, thought it should indeed be written, but did not have time to write it himself. Huxley was a busy and increasingly famous man and was never solely or even predominantly interested in racial matters. In 1941 he felt too busy to take on *Argument of Blood*, knew that his name would attract readers to the book and trusted Singer to actually do the work. It is feasible that the same could be said of *We Europeans*. Barkan is no doubt correct that some or all of those involved with the project thought it best to present the text as Haddon's and Huxley's for political reasons. However, the example of *Argument of Blood* suggests that another key reason for the extent of team writing in *We Europeans* may have been Huxley's celebrity and commitments.

As it stands, who actually wrote *We Europeans* will never be fully clear. Haddon claimed responsibility for Chapters 6 and 7, but only before noting that his writing had been subject to 'a certain amount of amendation'.[108] As we have seen above, Carr Saunders submitted Chapter 8 on 'Europe overseas'. In other sections, the tone of specific authors is detectable. Where *We Europeans* expressed the view that races could not be set up in a hierarchy of ability because of the extent to which they overlap with each other, one can detect the pen of Huxley, so similar is the argument to that which he made in *Africa View* in 1931. Comparing the following two extracts illustrates this point:

We Europeans:
>...mental differences must be expected to be like the physical in being mere matters of general averages and proportions of types – there will be in every social class or ethnic group a great quantitative range and great qualitative diversity of mental characters, and different groups will very largely overlap with each other.[109]

Africa View:
>...the difference between the racial averages will be small; and that they will only be an affair of averages, and that the great majority of the two populations will overlap as regards their innate intellectual capacities.[110]

If the above extracts reveal the contribution of Huxley, where *We Europeans* makes the more radical case, that race is a social construct and not a biological category, it seems possible to see the contribution of Charles Singer, who made this point forcefully in his correspondence throughout this period.[111]

Whilst there will remain doubt about who wrote what in *We Europeans*, there can be no doubt whatsoever about the political impact of the text. At the time of publication, *We Europeans* was heralded as an iconoclastic scientific statement, challenging the utility of the term 'race' in science and society. For example, sociologist Cedric Dover's review in *Nature* described the text as 'attractively radical', 'an opportune prophylactic against the spreading virus of racialism'.[112] The radicalism that Dover saw in *We Europeans* undoubtedly concerned its treatment of race. He explained:

> The main utility of the book is that it demonstrates with incontrovertible clarity (and it is a reflection on human sanity that such a demonstration should still be necessary) that the racial concept is a myth, and race 'a pseudo-scientific rather than a scientific term'.[113]

In many ways, this reading of *We Europeans* was justified. *We Europeans* attacked the inconsistent usage of the term 'race' in European discourse. It argued: 'The term "race" is freely employed in many kinds of literature, but investigation of the use of the word soon reveals that no exact meaning is, or perhaps can be, attached to it, as far as modern human aggregates are concerned.'[114] The case was made that not only was race as an idea used inconsistently, but that there was no justification for using it at all. *We Europeans* asserted that European populations were so racially mixed that 'nothing in the nature of "pure race" in the biological sense has any real existence'.[115] All modern populations were 'melting pots of race'.[116] In this context, *We Europeans* called for a change of vocabulary in scientific writing about different populations. In this new vocabulary there was to be no place at all for race. In recognition of the case made, that 'all that exists today is a number of arbitrary ethnic groups, intergrading into each other', the authors contended: 'For existing populations, the word *race* should be banished, and the descriptive and non-committal term *ethnic group* should be substituted.'[117]

A corollary of this argument was that there was no biological case for attempting to racially control European population mixing. Making a major challenge to conservative scientific thinking on 'miscegenation', *We Europeans* argued that the 'biologist and the eugenicist ... [had]...a negligible or at best a minor role to play as advisors' when it came to deciding who should mate with whom. When scientists became embroiled in racial mixing policy, the authors asserted, it was not out of any concern rooted in empirical fact but needed to be seen instead as the manipulation of scientific theory into a 'cloak to fling over obscure, perhaps unconscious feelings'.[118] This case about 'miscegenation' was written with one eye on Nazi Germany's growing raft of policies and proclamations about racial purity. In fact, *We Europeans'* general argument on race should be read as an explicitly political statement about Nazi Germany, a call to all scientists to join together and refute Nazi racial theory. 'Science must refuse', the authors argued by way of a conclusion, 'to lend her sanction to the absurdities and horrors perpetrated in her name. Racialism is a myth, and a dangerous myth at that.'[119]

This political goal was not lost on those scientists who were opposed to the idea of challenging race as a concept. For Reginald Gates, a friend of Huxley's since their association with the *Journal of Experimental Biology* in the early 1920s and through Huxley's time at King's College, *We Europeans* was nothing less than a betrayal. Though this chapter will go on to argue that the argument of the book was not nearly as radical on race as it may first appear, its anti-racial message was enough to infuriate Gates, who remained throughout his career a strong believer in the idea of race and a fierce opponent of 'miscegenation'.

Gates's anger about *We Europeans* was not only based in his hostility towards the ideas about race expressed in the book. He also felt that the book's argument was a breach of scientific integrity, a piece of political propaganda and not a work of science. In a series of acrimonious letters, Gates remonstrated with Huxley about the text, especially about the involvement of A.C. Haddon who, Gates was convinced, had been manipulated into serving as co-author. One letter from Gates in March 1937 summed up all his feelings about the *We Europeans* project:

> ...please don't pretend that Haddon agrees with you about the book. You must know that he tried to withdraw from it when he found what you were making of it, and that he is heartily disgusted with the whole thing. How you, knowing no anthropology, had the nerve to force an authority like Haddon to accept your biased and controversial statements passes my comprehension...In my opinion the book represents you at your worst. Such tendentiousness and one-sided statements bring both anthropology and genetics into disrepute; and the more so when they pose as 'an attempt at fair statement'. The book is, in my view, a clear case of misuse of your powers...I cannot remember ever reading a supposedly scientific book which gave me this feeling so strongly.[120]

A few aspects of this attack are particularly illuminating. There was some truth in Gates's accusations about Haddon's disaffection from *We*

Europeans. As Barkan has noted, Haddon was at best ambivalent towards the project, expressing conflicting opinions at different times.[121] In 1935, Haddon seemed happy enough with the text. Having expressed the reservation to Huxley in August that Singer and Seligman were 'bitter on the subject [of race] and might feel inclined to let themselves "go"', he wrote again a month later to say that Singer had 'greatly improved the book'. Haddon added: 'Really am pleased with the book – it is much needed – no more so than in Germany – where it will be banned.'[122] However, as time passed it seems that Haddon's enthusiasm for *We Europeans* waned. In a later letter from Gates to Huxley, Gates quoted a seemingly disillusioned Haddon who told him that he had 'tried to withdraw from the concern but was over-ruled' and that he was 'very sorry... [to have had]... anything to do with it'.[123]

Probing into the authorship of *We Europeans*, Gates, in the same letter to Huxley that is cited above, added that he knew 'a good deal of the history of *We Europeans*'. This was Gates's way of telling Huxley that he thought that he (and not Haddon) had written the bulk of the text. In another letter Gates expressed this view explicitly: 'The controversial parts of that book are so clearly contrary to his [Haddon's] method and so obviously written by you that I think it very unfortunate that you should attempt to draw him in to defend you in any controversy regarding your denial of race.'[124] Despite thinking that he knew who had written *We Europeans*, Gates actually had no idea. Had he known of the role of Singer and Seligman, given the ferocity and fanaticism of his conspiratorial anti-Semitism (of which we shall see much more below), his criticisms would have been tenfold. That Haddon did not feel the need to tell him perhaps finally settles his relationship to the project. Unsure as he may have been about *We Europeans*, Haddon was not willing to give Gates the ammunition he needed to blow the cover of the book's true authorship.

Ultimately, Haddon's ambivalence about *We Europeans* reflects an accurate criticism of Gates's; namely that the stance on race that was taken in the book was political, much more than it was scientific. As Haddon is recorded as noting above, the book was needed, especially in Germany, for political reasons. It seems likely that when challenged by Gates about the stance taken on race in *We Europeans*, Haddon was slightly ashamed and attempted to distance himself from it. There is nothing in the canon of Haddon's work or in his papers that in any way approximates to the determined challenge to race expressed in *We Europeans*, and it is unlikely that Haddon actually believed in this challenge from a scientific point of view.[125] His involvement was based instead on a willingness (albeit unsure) to make a political statement, for Haddon believed in the 'ends' of *We Europeans*, if not the 'means'. In this, as we shall see, he was far from alone.

Deconstructing *We Europeans*, it is possible to see great uncertainty and confusion surrounding its ostensible abandonment of race. Looking in detail at the text, it is arguable that in fact the critique made about race was far more limited and ambiguous than it first appears. For one thing, *We Europeans* did not go as far as to suggest that different populations were in any way equal and in fact defended the idea that racial mental differences existed between different peoples. The authors concluded (regarding the notion that some ethnic groups were mentally inferior): 'These objections undoubtedly have some validity.'[126]

The critique of race in *We Europeans* was grounded in scientific uncertainty. The point made was not that racial differences did not exist between populations, but that science at its present stage was not capable of providing any definitive judgements about race and its effects. Thus *We Europeans* contended not that all disparities between groups were cultural/environmental, but only that it was impossible to draw with any certainty a line between those differences which were cultural and those which were inherent. 'With the best will in the world...', Huxley and Haddon argued, '...it is, in the present state of knowledge, impossible to disentangle the genetic from the environmental factors in matters of "racial traits", "national character", and the like.'[127] The authors of *We Europeans* drew upon this uncertainty to argue for a moratorium on any statements about racial mental difference, until science could work the matter out:

> Until we have invented a method for distinguishing the effects of social environment from those of genetic constitution, we shall be wholly unable to say anything of the least scientific value on such vital topics as the possible genetic differences in intelligence, initiative and aptitude which may distinguish different human groups.[128]

As radical as this message may have been, set against the racial dogmatism of the Third Reich, it was by no means a refutation of race. *We Europeans*' authors were prepared to attack the fallacies of Nazi racial theory, very willing to stress the importance of environmental and cultural factors in shaping differences between groups, but could not bring themselves to dismiss the biological possibility of racial diversity.

Similarly, if we explore in detail the consideration given in *We Europeans* to the issue of 'miscegenation' it is again possible to see an authorial unwillingness to abandon altogether the idea of racial difference. Whilst, as has been noted, the authors of *We Europeans* made a determined attack on the factual accuracy of the scholarship of those who railed against 'miscegenation', they were ultimately unwilling to confirm that such mixing would not lead to racial damage. Referring to the 'question of the biological results of wide crosses' (which should be read as a euphemism for the sexual mixing of white Europeans with black non-Europeans) the authors recorded that it was 'extremely difficult to come to any firm conclusion' as to the effects. Nonetheless, *We Europeans* did entertain the idea that 'very wide crosses may give biologically "disharmonic" results in later generations, by producing ill-assorted combinations of characters'.[129] Whilst asserting that issues of race mixture were not primarily about race but 'nationality, class or social status', *We Europeans* did not argue that there was nothing wrong with 'miscegenation'. In fact the authors concluded that whilst the racial mixing of wide crosses 'might conceivably be a good thing', there was 'a limit to the amount of foreign stock which can be taken up by a nation in a given time' and that '"racial crossing" may be inadvisable, but chiefly because the ethnic groups involved happen to be in different national worlds or different cultural levels'.[130] This ambivalence was not lost on reviewers. Cedric Dover, in *Nature*, criticised the weakness of *We Europeans*'s stance on 'miscegenation'. He noted dryly: 'It rather seems as if Dr Huxley suddenly remembered the manifesto of the Eugenics Society and his position on its council.'[131]

We Europeans undoubtedly served as a foil to Nazi racial theory and posed a wider challenge to the racial views of the scientific establishment in Britain. Its message was radical for the mid 1930s, though not as radical as it may appear at first glance. As such it provides a window into the progressive challenge to racial biology in the period. For just as *We Europeans* struggled to dismiss the idea of mental racial difference and the dangers of miscegenation, Britain's leading progressive biologists were similarly uncertain about the racial challenge that they were making. However, just as in *We Europeans*, a challenge was made to conservative racial science, driven by a political will if not full scientific conviction.

The limits of racial reform before the Second World War

The contradictions and confusion of racial reform in this period often emerge in a disparity between what scientists wrote about race when they perceived that there was an enemy to be challenged (particularly Nazi Germany) and what they actually thought. Leading scientific luminaries like Julian Huxley, J.B.S. Haldane and Lancelot Hogben all went out of their way to put over anti-racial messages in this period, messages that were often not fully replicated in their personal correspondence and wider writing on the subject. Indeed, the racial views of the progressive biologists are harder to pin down than one might think having looked at their popular writing.[132] Their uncertainty and hesitancy have led to some scholarly confusion about race in interwar British science, especially concerning the racial outlook of Britain's growing Eugenics Society.

Under the leadership of C.P. Blacker, Britain's Eugenics Society, always a broad church, veered on race towards the perspective of the progressives. This tendency was part of a wider ideological journey within the interwar Eugenics Society, characterised by Kevles and Soloway as a passage from 'mainline' to 'reform' eugenics.[133] The platform of reform eugenicists differed from that of their mainline predecessors in several respects, encompassing both scientific and political shifts in aims. The most notable issues were not racial ones; the reform agenda worked primarily to ameliorate the class biases which had characterised earlier eugenic thinking and recognise the importance of environmental (as well as hereditary) factors in shaping social development.[134]

This change in direction within the Society can be seen as a political attempt to maximise its appeal and influence and ensure that it did not alienate the political left (in science or society) by stigmatising the qualities of the British working classes. However, it carried with it attendant implications for the Society's stance on racial matters. Fuelled by a desire to keep the left on board, and driven by a personal liberal outlook, Blacker worked hard to ensure that British eugenics maintained a moderate tone on race, and to protect the Society from harmful comparisons to the rightist extremism of Nazi Germany.[135] Thus, the Eugenics Society in the 1930s and 1940s avoided campaigning on the old stalwarts of working-class degeneracy and immigration, focusing instead on ostensibly less controversial issues, mostly related to broader questions of population development. These included the need to develop voluntary birth control options for families and (although it may seem contradictory) the importance of sustaining the population of Britain. Blacker's Society thus cast itself as the natural home of the science of demographics. Financially buoyed by a substantial bequest in 1930, the Society brought together and funded a range of research bodies to explore these issues within a mostly liberal and scholarly atmosphere.[136]

Accurately positioning Britain's leading progressive biologists in Blacker's reformist Eugenics Society is not an easy business and has led to some historical confusion. For example, Nancy Stepan, having written that 'Hogben, Haldane and Huxley all found themselves in the anti-eugenic camp by the 1930s', almost immediately has to concede the continuing involvement of Huxley and Haldane in the movement through the decade.[137] In fact, Huxley remained a card-carrying eugenicist throughout his career, became President of the Society in the 1950s and gave the prestigious annual Galton lecture as late as 1962.[138] Even Lancelot Hogben, the most racially radical of the three biologists, and a firm objector to mainline eugenics because of its anti-working-class tendencies, was not entirely unsusceptible to Blacker's reformist wooing.[139]

Despite having asked for his name to be removed from the list of the Eugenics Society's Consultative Committee in 1933, Hogben served on several other bodies which considered population questions on behalf of the Society.[140] This behaviour reveals Hogben's true stance towards eugenics which was ambiguous more than it was consistently hostile. In 1935, Hogben wrote of 'the necessity of laying firmly the foundations of an exact science of human inheritance', an agenda which would have won the support of any eugenicist of the day.[141] Indeed, his ambivalence and hesitancy about the discipline were interpreted by C.P. Blacker as a beneficial contribution to British eugenics. Blacker said of Hogben's work in the 1945 Galton lecture: 'It is my opinion that Professor Hogben's criticisms have been useful to the society, and I, for one, feel indebted to him, though he has made it sensibly more difficult to give a simple lecture on eugenics.'[142]

Blacker's warmth towards Hogben's criticisms of eugenics reflects both his determination to drive the Eugenics Society into a liberal stance on racial matters and something of the continuing interests and values that were shared between the Eugenics Society and leading progressives. Although it would be inappropriate to mask growing conflicts over class and race where they existed, it is beyond doubt that Hogben, Haldane, Huxley and Crew all worked with Blacker on eugenical projects in this period, notably in the Birth Control Investigation Committee (BCIC).[143] So if Stepan's analysis of the relationship between the progressives and the Eugenics movement is somewhat confused, that is because the matter is very confusing, not least because scholars like Huxley, Haldane and Hogben were themselves increasingly perplexed by the question of race.

As the ambiguities in *We Europeans* suggest, Huxley's views on race remained complicated in this period. Despite his determined and unequivocal opposition to Nazi racial theory, Huxley did not feel the need to completely overhaul his own racial views. This stance is evident in an anti-Nazi article that he wrote for an American audience,

'Aryan Racial Myth Exploded', published in the *San Francisco Examiner* in October 1938. In this piece, Huxley as usual attacked the Nazi idea of racial purity, noting that 'every so-called race is largely hybrid' and concluding that 'the whole of racialism as a doctrine, whether Nazi or Fascist, whether British or American, is a myth...[that]... becomes particularly dangerous when it receives official support and becomes sacrosanct orthodoxy'.[144] However, in this same article Huxley made forceful racial claims of his own. In a passage which is worth citing at length, he not only argued that peoples differed as regards both physical and psychological qualities, but actually expressed these differences in racial terms by noting that traits were orientated around 'the primary races of man'. Huxley even went as far here as to suggest white superiority, dismissing the racial achievements of 'Negro' and 'American Indian' communities:

> There is first the undoubted fact we have mentioned, that different peoples are not alike. There are physical and temperamental differences. There are also differences in achievement. No Negro people ever invented ploughing, written language or stone architecture. White men are the most advanced in science and material civilisation (though the Japanese are threatening their supremacy). The American Indians do not rival the white invaders of their countries...It is probable that the primary races of man – black, white, yellow and brown – do differ inevitably and hereditarily, not only in their skin colour and other physical characters, but also in their temperament and their intellectual capacities. These differences, of course, will only be average ones; individually there is an enormous overlap in all mental qualities.[145]

This article seems to sum up Huxley's perspective on race and shows his true colours. Whilst no one can doubt his hostility towards Nazism or question his determination to oppose Hitler's construction of pure European races, this commitment did not add up to a dismissal of race as an idea.

Huxley's ambivalence in many ways sums up the progressive biologists' perspective. In the same way that Davies and Hughes and Caradog Jones challenged the racial analysis of their predecessors by conditioning, not dismissing, the idea of inherent racial difference, so Huxley and other leading progressives challenged the conservative scientific establishment not through radical anti-racial analysis but by seeking out a middle ground. In 1936 Huxley gave the prestigious annual Galton lecture to the Eugenics Society, a lecture which nicely indicates the changing reformist tone of British eugenics and the relationship of leading racial progressives to the Society. Huxley used much of the lecture to lay out his 'middle ground' approach to racial study. Having registered his disgust with the bad racial science of Nazi Germany and his belief that the term 'race' was now too problematic to use, Huxley went on to outline a firm belief in physical and mental racial differences between populations.[146]

> There is no doubt that genetic differences of temperament, including tendencies to social or anti-social action, to cooperation or individualism, do exist, nor that they could be bred for in man as man has bred for tameness and other temperamental traits in many domestic animals, and it is extremely important to do so....The variability of man, due to recombination between divergent types that have failed to become separated as species, is greater than that of any wild animal.[147]

In firmly emphasising his ongoing belief in this kind of difference Huxley dismissed the 'sentimental environmentalism' that was to his mind becoming more commonplace in scientific challenges to Nazism, and instead proposed a middle way.[148] Like Davies and Hughes, he emphasised the importance of environmental factors in shaping individual and group behaviour as a partner to heredity. This was really the key point of the lecture, as Huxley told the Eugenics Society that it was necessary to understand both environmental and hereditary factors in order to explain human difference. He argued: 'Any character whatsoever can only be a resultant between genes and environment...In general, neither nature or nurture can be more important, because they are both essential.'[149] In recognition of the importance of both heredity and environment Huxley called for cooperation between sociologists and eugenicists, indicating his continuing belief in racial difference but also his recognition that many so-called racial problems had their roots in politics and society and not in natural science.[150]

That Huxley's case represented a middle ground in science is evident in some of the responses to his Galton lecture. For some, like T. Drummond Shields from the British Social Hygiene Council, Huxley's lecture represented thoughtful, progressive eugenic opinion. Drummond Shields wrote to Huxley:

> I am afraid that the status and authority of the Eugenics message has been adversely affected by its being mixed up with a

> good deal of social snobbery and it is most valuable to have had that corrected by such a very comprehensive and able address.[151]

For others, however, Huxley had not gone nearly far enough in his critique of conservative eugenics. E.S. Goodrich, from the Department of Zoology and Comparative Anatomy at the Oxford Museum, berated Huxley for continuing to assert the eugenic opinion that racial qualities could be bred for.

> By the way I was sorry to see that in your Galton lecture you apparently favoured Muller's revolting suggestions. Is the fate of the human race to be put in the hands of a few individuals – probably extremists and fanatics! Are Germans to be bred according to notions of most aggressive Nazis? Is Italy to be flooded with the sperm of Mussolini? The whole idea seems to me most dangerous and unpleasant.[152]

Neither really radical nor conservative, Huxley's call for the continuation of a reformist eugenic agenda which recognised the horrors of Nazism and the importance of environment, epitomised the progressive challenge to race before the war and the evolving character of the British Eugenics movement.

A similar analysis can be made of J.B.S. Haldane. Haldane like Huxley was a determined opponent of Nazism, and a strong critic of Nazi racial policy. Haldane dismissed Nazi thinking on racial separation as 'completely meaningless' and contended that Nazi laws on racial purity had no basis in scientific fact.[153] However, like Huxley, Haldane still felt that race was important and that it would be a gross scientific error to abandon analysis of the subject.

Haldane expressed these views in some of his published writing in the interwar period. In his 1938 text *Heredity and Politics*, Haldane argued forcefully that there was psychological racial diversity between populations: 'there is no question in my mind of the existence of racial differences in psychology, but I do not know to what they are due'.[154]

Ultimately, Haldane's work sits firmly in the progressive camp precisely because he periodically expressed ambiguous opinions about racial difference. Haldane's conclusion in *Heredity and Politics* is revealing:

> I can only close this question of the alleged superiority of whites over Negroes on a note of agnosticism. I can state the view that not

merely has nothing been proved, but that it is going to be exceedingly difficult to prove anything within the next two generations.[155]

Haldane's uncertainty epitomised the British progressive biologists' perspective, where a critique of racial thinking was not launched from a viewpoint of unequivocal animosity to racial categorisation or to eugenics but in hostility to the abuse of racial theory by governments and a degree of scientific timidity about drawing firm conclusions regarding the roots of racial difference. Huxley and Haddon's acceptance that there was 'some validity' in racial categorisations should be seen, like Haldane's perspective, as an important departure from more extreme racial thinking.

As we have seen from *We Europeans*, those scientists who were moving away from racial thinking in this period were mostly not yet ready to disavow decisively the harmful effects of interracial mixing, especially between black and white people. As Haldane noted in one 1934 paper: 'A good case can be made out for discouraging immigration of Negroes into Europe, or of Europeans into tropical Africa, since in each case the immigrants are ill adapted.'[156] However, progressives like Haldane and Huxley were much more ambivalent and calm regarding the prospect of 'miscegenation' than many conservative racial scientists. For example, the analysis in *We Europeans*, which held that it was 'extremely difficult to come to any firm conclusion' as regards the effect of racial mixing, reflected wider progressive opinion and stands out sharply from conservative thinking on the subject.[157] Understanding the views of conservative biologists and physical anthropologists on race and racial mixing illuminates clearly the challenge made by the progressive biologists in the 1930s.

The leading defenders of race in British science were far less ideologically inclined to challenge the Third Reich than the progressives. Views on Nazism and Fascism ranged amongst conservatives between ambivalence to overt support. The 1930s writing of Britain's most senior racial conservative, Sir Arthur Keith, provides the most important and obvious example of this different stance. Keith was convinced that Hitler was generally correct in his approach to European politics and argued that the Nazi regime was right to develop policy that was wedded to the principles of scientific racial theory.

At the core of Keith's beliefs was the idea that nations were evolving races, competing in a faux Darwinian way in the battle for natural selection. He outlined in a 1928 lecture that 'nationalities are real races in a microdiacritic stage of evolution'.[158] Keith saw the First World War as the harbinger of a new age of racial/national commitment: '...the Great War has swept Europe, uncovering primitive traits and impulses which are deeply buried in human nature'.[159] The corollary of this opinion in evolving national feeling was the belief that racial prejudice was natural behaviour, biologically rooted in the make-up of the human psyche as a defence mechanism.

> Race-feeling lies latent in men and women as long as they move among their own kind; but when they move outside the frontiers of their tribe or country deep instinctive feelings of race-prejudice are awakened, and under certain circumstances may become inflamed or uncontrollable.[160]

Keith saw this natural racism as a benevolent force which would serve to develop the strengths of the human species. In a 1931 book, *The Place of Prejudice in Modern Civilisation*, Keith concluded: 'this antipathy or race prejudice Nature has implanted within you for her own ends – the improvement of mankind through racial differentiation. Race prejudice, I believe, works for the ultimate good of mankind.'[161]

These views reflected Keith's political sympathy for precisely those causes which drove the progressive scientists to make challenges to racial theory; he believed both in colonialism and Nazism. In a 1932 lecture Keith told a meeting of the National Union of Students that British 'efforts at colonisation' were 'justified' and represented 'the soundest and boldest attempts ever made to rationalise the total population of the earth'. Showing something of the Machiavellian political outlook that would enable him to express support for Hitler's race policy, Keith argued that even if colonisation led to 'the eliminations of native peoples, it is perhaps the least cruel of all the methods which aim at establishing peace and industry over wide regions of the world'.[162]

As late as 1941, Keith was willing to defend the racial goals of Nazism. In the context of war, he managed to express his patriotism without abandoning his support for Hitler's racial policies by arguing that his understanding of the logics of the Third Reich was important to the war effort. Giving a lecture entitled 'Race and Propaganda', Keith contended: 'At the risk of being called a traitor I maintain that Hitler is right in his conception of race and that my British colleagues are wrong. In a matter of life and death it is fatal to make a mistake about your enemy.'[163] Keith argued that his understanding did not equate to support. The fact that he appreciated that the Third Reich was trying to fulfil its national and racial destiny did not mean that he thought that

Britain (on her own racial quest for power) should allow her to do so. He concluded the lecture by arguing: 'Germany's racial pride is the supreme danger to the peace of Europe and of the world. No kind of paper pact will ever bind racial pride.'[164] Despite his careful positioning, in reality Keith was a Nazi sympathiser as were other leading racial conservatives like Gates and Pitt-Rivers.[165] This sympathy did not mean necessarily that these conservatives wanted to see Germany defeat Britain in war, but that they wished that the war had not occurred and felt that Britain's real enemy was the ostensibly anti-colonial and anti-racial Soviet Union.

Conservative views on race were driven by an unequivocal hostility to racial mixing which stands out sharply from the uncertainty of the progressives. To these scholars, British racial qualities were under threat and needed defending at all costs. Gates sustained the belief throughout this period that mixing between human races would lead to both mental and physical disabilities in the offspring, views similar to those that we have already seen expressed in Fletcher's report on mixed race children in Liverpool.[166] In a similar vein, University of Glasgow geologist J.W. Gregory concurred that 'miscegenation' would lead to a 'chaotic constitution' in children.[167] 'The result', he concluded, would be 'invariably a bad one.'[168] In an argument that would have sat well with Nazi racial analysis, teacher, writer and eugenicist Charles Wicksteed Armstrong argued that for Britain, racial mixing was 'throwing away her most precious heritage, more precious by far than empire or art or literature – the purity of her blood'.[169] Whilst Armstrong was a marginal figure, the views he expressed were not uncommon. As well as being the mainstay of Gates's and Keith's science, similar opinions on the physical ill effects of racial mixing were routinely expressed by numerous eugenicists in this period: Crew, Ludovici and Aikman amongst others.[170]

The continuation of this conservatism concerning racial type and mixing into the Nazi period can only be understood with reference to politics. In terms of science there really was little to divide Britain's leading biologists. Science had united Huxley, Haldane, Gates and Crew in the *Journal of Experimental Biology* in 1923. In 1933 it did not divide them. What actually drove Gates away from the rest was his politics. Faced with the horrors of the Third Reich, most scientists began to challenge racial theory, not because they did not believe in race, but because they felt a political need to do so. Right-wing scholars without the inclination to challenge Hitler were left behind because of their politics and not because of their science, a point illustrated by a brief return to the acrimonious correspondence between Gates and Huxley in 1937. Arguing with Huxley about *We Europeans*, Gates complained bitterly about comments that Huxley had made about him in his 1936 Galton lecture, which had criticised his views on race as expressed in the 'Race and Culture' conference.[171]

> At the Royal Institution you singled me out for attack, although Haldane's views, in the Report from which you quoted, were very similar to mine, and any differences expressed were quite negligible. Indeed, Haldane is on record as affirming (contrary to your expressed views) that the main types of living man must be recognised as distinct races. So I suggest that the next time you want to attack someone in the interest of your views, you attack him for a change.[172]

Gates had a point. There was very little scientific distance between his own racial views and Haldane's, not only during the 'Race and Culture' conference but in general.[173] Gates (and racial science) came under attack in this period because of political ideas and not because there was anything much different between their science and that of the progressives. That the bulk of British scientists began to rethink their racial views in the 1930s had everything to do with their ideological inclinations and little to do with science itself. In a period where British scientists felt increasingly able and inclined to intervene in the political arena, rather minor scientific differences were utilised in an attempt to reset the racial views of the nation.

Science and society in the 1930s

In a perceptive analysis of interwar biology, Roger Smith has argued that the 1930s witnessed an increasing closeness between scientists and the general public, a new level of unity which he describes as 'a shared and durable world of expression and judgement'.[174] As evidence for this increased proximity, Smith cites the development of the British Association for the Advancement of Science in this period and the rise of popular science publishing, especially by Penguin.[175]

There is strong evidence to support Smith's analysis. For one thing, leading biologists repeatedly attempted in this period to bring science to the public through ambitious accessible texts; Huxley and Haddon's *We Europeans* is an example of one such offering. As we have seen, this book was explicitly aimed at a wide social readership, not at all written for scientific peers but for the people. Similarly ambitious projects were fairly commonplace. Most notably, Lancelot Hogben's *Science for the Citizen* and Huxley's collaboration with H.G. and G.P. Wells, *The Science of Life*, stand out as bold attempts to bring the general public into the scientific community.[176] It seems reasonable to view the writing of these texts and countless others as a conscious attempt by some of Britain's leading luminaries to democratise science and develop the 'common man'. Hogben argued at the beginning of *Science for the Citizen*: 'Natural science is an essential part of the education of a citizen, because scientific discoveries affect the lives of everyone.'[177] The democratising mission was of central importance to new Marxists like J.B.S. Haldane who wrote science columns for (and at one point edited) the *Daily Worker*.[178] Describing the agenda of leftist scientists in this period, Werskey has recorded the centrality of popular dissemination: 'No opportunity or audience would be missed in their doubled-barrelled efforts to enlighten the world about the scientific dimensions of socialism and the social relations of science.'[179]

However, if would be a mistake to view the popularisation of science in the interwar period as merely the concern of leftists. The desire to bring science to ordinary people reflected a wider progressive scientific tendency, rooted in the belief that society could and should be developed on rational and scientific lines. As Haldane himself argued long before his Marxist rebirth: 'If we are to control our own and one another's actions as we are learning to control nature, the scientific point of view must come out of the laboratory and be applied to the events of daily life.'[180]

The political context was far from irrelevant in persuading scientists in this period to reach out to the general population. Just as *We Europeans* was designed to prevent the spread of Nazi-style racism, other key works were similarly conceived as weapons against the dictators. Hogben wrote *Science for the Citizen* in the belief that only the diffusion of scientific rationalism could save the European citizen from being seduced by extremism (as he perceived had happened in Germany). Hogben felt the need to respond to a European society where only the dictators were offering any kind of future for the common man. In the epilogue of *Science for the Citizen*, he concluded:

> The revolt against the beehive city of competitive industrialism has already become a retreat into barbarism; and the retreat will continue unless science can foster a lively recognition of the positive achievements of civilisation by reinstating faith in a future of constructive effort.[181]

In Hogben's mind, the democracies needed to match the energy of the dictators with practical scientific development to ameliorate the lot of their citizens. As he noted: 'This is not the age of the pamphleteers. It is the age of the engineers...Democracy will not be salvaged by men who can talk fluently, debate forcefully, and quote aptly.'[182]

It was in the context of this perceived need for action that the progressive luminaries of British biology reached out of their laboratories to a more general audience, no one more so than Julian Huxley.[183] In an undated paper provisionally titled 'Science and Synthesis' Huxley outlined the need for scientific engagement with world affairs and with it the rationale which seems to have underpinned his entire career.

> Any scientist who emerges from the safe rabbit-burrow of his special subject and takes a good look at the world's general situation is virtually forced to assume the mantle of a prophet. He prophesies doom; but, being a scientist, the doom he prophesies is a conditional doom: he prophesies doom unless we do certain things; if we do them, doom is avoided, and he can become a prophet of destiny.[184]

Huxley's election to the Royal Society in 1938 can be seen as a coming of age for this outward-looking style of science. Having previously been passed over by the Society, seemingly because of his populist approach, Huxley's belated election triggered an emphatic response from peers and supporters. H.J. Fleure wrote to Huxley: 'Your championship of science in our social life seems to me to be a very valuable thing. I hope you may be helped by this election to be still more effective.' Similarly, in Charles Singer's letter of congratulation to his friend he noted: 'The cause for delay is clear enough, but I hope you will continue without hesitation to give some part of your time to human as distinct from scientific matters.'[185] The Royal Society, which had previously looked with a degree of condescension at Huxley's populist writing, was by 1938 prepared to recognise the growing influence of his approach. At the highest levels of academia, by the late 1930s, science was increasingly ready to stake its claim on the time and policies of wider society.

Whilst the popular writing of Hogben, Huxley and others may have brought science to a larger audience than ever in the 1930s, it has already been noted that scientists with an interest in race had been trying to stake their claim on the imaginations of policy makers and the public throughout the interwar period. This chapter has recorded the way in which scientific and quasi-scientific groupings like the Jewish Health Organisation of Great Britain and the Liverpool Association for the Welfare of Half-Caste Children commissioned scientific research in order to lobby in favour of particular immigration and integration policies. In addition to these organisations, the Eugenics Society worked tirelessly to convince policy makers and the public of the sagacity of its views on Britain's racial development throughout the interwar period.

Mazumdar explains eugenics in the context of its goals beyond the scientific arena, as an attempt 'to apply an understanding of the laws of biology to the laws that determined the lives and environments of the subjects of the realm'.[186] Above and beyond abstract discussion of racial advancement, the Eugenics Society existed in order to campaign for those changes in social policies and habits which its members perceived would improve the racial stock of the nation. C.P. Blacker noted as much in his 1945 Galton lecture where he considered the role of the Eugenics Society: 'Our primary object is not the glorification of the Society, nor even the glorification of the word eugenics; it is to get sound eugenic principles recognized, accepted and acted upon.'[187] Towards this goal, Blacker's Society had some success, significantly influencing government papers in the interwar period on mental deficiency and the potential need for voluntary sterilisation.[188]

Mazumdar has outlined that the social concerns of the Eugenics Society were predominately focused on 'poverty and pauperism'.[189] However, Dan Stone's argument, that the Society took race (and not just class) seriously, has been an important challenge to Mazumdar's generally sound analysis. As well as noting the publication of articles on race in *Eugenics Review*, Stone has cited the research by Rachel Fleming into Anglo-Chinese race crossing (mentioned above as a forerunner to the Fletcher report) and the interest shown by the Society in Caribbean immigration into Britain in the 1950s as examples of a continued eugenical interest in race.[190] It is fair to say that compared to the obvious and emphatic propagandist intentions of *We Europeans* and other anti-Nazi publications, the muted racial concerns of the Eugenics Society were perhaps not as pressing on the public's attention. They did however contribute to a cultural climate where scientists felt a desire and responsibility to try to influence and educate the public on race.

Smith's analysis of this process of influence, specifically of the diffusion of racial ideas from the eugenics movement to the public, merits our attention. It is rooted, as was argued in the Introduction, in his understanding of a 'common context' between science and wider society, an idea which implies a passage of influence and knowledge that operates in both directions between scientific experts and lay people.[191]

This denial of a scientific lead role in opinion shaping concurs with leading scholarship of scientific racism in Britain, which has repeatedly asserted that scientific writing on race cannot be held to have had any specific or direct impact on British policy and values in the 1930s, despite the best laid plans of many scientists.[192] If these scholars are correct and it cannot be argued that policy and attitudes were driven by racial science, then Smith's idea of a common scientific and social context on race is an appealing one. Using this framework of analysis in the context of this study, it could be argued that instead of leading society on racial ideology, science in fact operated in a shared discursive terrain as the key scientific ideas that dominated biological writing on race in the 1930s mirrored wider social racial thinking.

This chapter began with the contention that scientific racial debate in this period was focused around two core questions, one about the nature of population difference, the other about the effect of racial mixing. It has concluded that both progressive and conservative writers on race generally continued to share beliefs about the answers to these questions, notably that there were both physical and psychological differences between populations and that racial mixing would lead to uncertain results. It has argued that in the context of this continuing agreement, the progressive challenge can only be understood in political terms, as a willingness amongst some to moderate and mitigate racial beliefs in order to attack Nazi and other oppressive racial theories. Acknowledging that the progressives were firmly in the ascendancy by the beginning of the Second World War, it is possible to sum up most scientific perspectives on race in the 1930s in these three statements:

1. There are physical and psychological differences between populations, rooted in both hereditary and environmental causes.
2. Racial mixing may produce positive or negative results, but cannot be said to be unconditionally desirable or unproblematic.
3. Nazi racial policy is brutal and morally abhorrent.

In order to see the extent to which these core ideas reflected or shaped public and state understandings of racial difference, this chapter will now turn to a specific case study, focusing on government policy and social attitudes towards Jewish refugees from Nazism. This is an issue which, perhaps more than any other, brought discussions of racial difference into the non-scientific arena in the 1930s. By exploring this key issue of immigration and integration, this case study will enable a more detailed analysis of the idea of a public and scientific 'common context' on racial matters. As one prominent historian of Anglo-Jewish relations has put it, policy towards the 80,000 Jews that came to Britain between 1933 and 1939 can be read as 'an expression of the values of the society that produced it'.[193]

Refuge, race and restriction: Britain and the Jewish refugees from Nazism

The ideas which shaped responses to refugees in the 1930s were mostly constructed outside the arena of academic research.[194] On a scale which reflects but also surpasses the contradictory currents of British biology, the theories of race and racial difference that helped to underpin social reactions to refugees in this period were subjective, malleable and contested. Due to the depth of differences of opinion and tone, theories about Jewish racial stock or type were never static. In his seminal analysis of British constructions of 'the Jew', Cheyette has coined the term 'Semitic discourse' to describe the prevalence of various autonomous, often contradictory, racial Jewish images in the British imagination.[195] This chapter will utilise Cheyette's idea of Semitic discourse in order to explain some of the prominent racial images that were tagged to Jewish refugees in 1930s Britain and to consider the role of scientists in their generation and development. In this context three twentieth-century themes of Semitic racial discourse have been selected for analysis, highlighting perceptions of Jews as cowards, conspirators and disease carriers. This chapter will briefly outline these key themes, before going on to consider their impact.

Fin de siècle commentators often dismissed the Jew as a racial coward, untouched by the concepts of chivalry and valour. Eugenics enthusiast and writer Arnold White argued that if a conflict occurred in a country where Jews were resident, they would 'bring dishonour upon their race in times of war for the sake of gain'.[196] White based this slur on his analysis of Jewish behaviour in the Boer War: 'The foreign Jew in smart society did not go to the war', he recorded.[197] In White's opinion the Jew was inherently mercenary and totally bereft of the concept of patriotic bravery.[198] With ugly characterisation of supposedly Jewish dialect, White recorded the Jewish reaction to British reverses in the Boer War, as he perceived it:

> 'I like dis news; it vill gif a goot shake out to shtocks – dat iss healthy'. While the bodies of Anglo-Saxon and of Celt were lying

unburied on the veldt under the African sun, this Teuton Semite philosopher could see no other aspect of the reverse to our arms than it would 'give a good shake out to stocks'.[199]

In the wake of the First World War, the characterisation of Jews as unpatriotic cowards became a leitmotif of pro-restriction, anti-alien politics and agitation. As Jewish refugees sought a place in interwar British society, one of the biggest barriers to acceptance was the popular belief that Jews had shied away from 'doing their duty' in the Great War.[200]

Jews served in significant numbers in all avenues of the conflict. Kadish has noted that whilst 11.5 per cent of the nation at large went to the war, 14 per cent of British Jews saw active service.[201] This service, however, did little to moderate the widespread allegations of poor conduct levelled against British Jewry both during and after the war. The prominence of eugenicists in the evolution of wider public thinking regarding Jewish racial cowardice is notable. For example, Moul and Pearson mentioned this in their report on immigrant Jewish children as they criticised the pusillanimity of Jews who hid in underground train stations during First World War air raids. Whilst acknowledging that 'some' Jews behaved bravely in the war, the authors noted:

> Those of us who had occasion to travel during air raids on London will not lightly forget the sights and sounds we encountered among the Yiddish speaking population who sought refuge in the tube stations.[202]

More extreme eugenicist and anti-Semite G.P. Mudge, writing in the *Eugenics Review,* criticised Jewry in general on similar lines, yet with added virulence and anger.

> Our race played the game while these immigrants fattened in safety and under a double protection. In their frenzied efforts during the air raids to preserve their own skins, they frantically pushed through women and children, severely crushing them and even trampling upon them if they fell. These fine specimens of immigrant manhood cared not who sank so long as they reached shelter first.[203]

Eugenicist and medical practitioner Frederick Parkes Weber kept case notes on soldiers and civilians that he had treated during the First World War. His analysis of meetings with some Jewish patients revealed his strong belief in the idea of inherent Jewish cowardice. Parkes Weber recorded two cases of soldiers 'of Hebrew origin' suffering from shell shock.[204] In both cases, he clearly felt that the men were lying about their condition in order to evade service.

> After the war the condition improved. Once I met the patient's wife and asked about her husband, and she told me that he had so greatly improved that recently he had run off with another woman.[205]

> He understands that if he does not improve, he will be sent to another hospital – perhaps to an infirmary (which he will not like)…The house-physician suspects that patient had some interest (to avoid military service or something of the kind?) in making himself appear worse than he was.[206]

Views on Jewish cowardice were far from the preserve of eugenicists and were in fact common currency in wider British society. Holmes has argued that perceptions of poor Jewish war conduct made significant inroads into public opinion, which 'was seriously inflamed at an early date over the conscription issue'.[207] Jewish interest publications in the period repeatedly defended the service records of their community. *The Jewish Chronicle* met the accusations head-on in a 1919 editorial:

> The East End was a veritable hive of sustained war effort during the whole of the hostilities. Nor was it only a matter of munition making. Thousands of the children of Polish and Russian Jews fought manfully with the colours and gave their quota, alas! To the casualty lists.[208]

However, the perception of Jewry avoiding military service became so prominent and powerful as to lead the war government into legislation aimed at forcing active service on more Jewish immigrants. The Anglo-Russian Military Service Agreement of July 1917 ordered Russian immigrant Jews to choose between fighting for the British or returning to fight with the Russian army.[209] The immigrant Jews in question were recent refugees of often savage Russian persecution, but instead of perceiving them as victims in this way, it seems that the government was convinced instead by racial characterisations of Jews as cowards and draft-dodgers.[210] In this context, no allowance was made in recognition that it may have been difficult for refugees to fight on the side of the country which had so recently robbed them of their homes and communities.[211] As we shall see, similarly constructed thinking also seriously impacted on British refugee policy in the face of Nazism.

Many fringe and some mainstream scholars of race also addressed the idea of an inherent Jewish conspiratorial nature in this period. Parkes Weber and Gates amongst others expressed views concerning conspiratorial Jewish behaviour whilst the ever melodramatic teacher and eugenicist Wicksteed Armstrong argued (concerning the aims of Russian Jewish refugees in Europe): 'Unless the Western nations can unite in face of the common peril, the result may be, in the near future, such slaughter as will pale the Great War into mere child's play, and throw back human evolution a thousand years.'[212]

In the insecure climate of the interwar years, these images fuelled ideas of Jewish conspiracy against Britain. Popular allegations held that Jews were responsible for international problems (usually Bolshevism and German militarism) and that Jews had profited out of the war whilst the rest of the nation had suffered. It is possible to see the prevalence of this kind of thinking in state policy, as beliefs in Jewish war gain resonated around the Houses of Parliament during the 1919 debate on the Aliens Restriction Amendment Bill. A rhetorical question in the Commons from Conservative MP Sir John Butcher asked if it was good for the country to tolerate 'alien' communities:

> ...to interfere with our own people, to set up the same system of intrigue in our midst, the same system of interference with British Labour, the same system of undermining British business that we had before the war?[213]

Another member argued that these immigrants had 'always been traitors to the British workers as well as traitors to the British cause...Bolshevism, [was] of course...introduced in England almost entirely by aliens'.[214] Clearly, many MPs also believed that Jewish immigrants had higher criminal tendencies than indigenous Britons. Extreme anti-Semite Ernst Wild told the House: 'You cannot be in the criminal courts without realising what an enormous amount of the work of our courts is caused by the aliens and their crimes.'[215]

Jewish immigrants were similarly often characterised as being dangerous in terms of the risk they posed to the physical health of the nation. Sometimes, analysts used allegations of hereditary proneness to disease as proof of the inferiority of the Jewish 'type' (as Moul and Pearson did in their report into the comparative health of immigrant

Jewish children). The notion of Jewish proneness to illness often fed into wider characterisations of Jews as defilers and corrupters of other peoples. These perceptions were outlined in pro-restriction agitation around the turn of the century. In 1902 Dr Francis Tyrell informed the Royal Commission on Alien Immigration that Jews were uniquely susceptible to chronic granular ophthalmia and that Jews would spread this disease among the indigenous population.[216] Similarly, MP Evans-Gordon told the Commons in 1905 that smallpox, scarlet fever, trachoma and favus had undoubtedly been introduced by aliens.[217] Evans-Gordon described his contact with arriving aliens in graphic terms:

> We found some of them suffering from loathsome and unmentionable diseases, the importation of which into this country might and does lead to very serious results, and we found most of them verminous.[218]

A generation later, Parkes Weber asserted that Jewish families had 'special tendencies to certain metabolic (notably obesity and diabetes mellitus) and neurological (including psycho-neurotic) diseases'.[219]

Thinking on Jewish conspiratorial tendencies also convinced some theorists that Jews retained a special immunity to various diseases. Gates certainly believed that Jews had special powers to protect themselves both from disease and attack. His notes on the subject are revealing: '[Jews are]...Not affected by great epidemics of Mid[dle] Ages in Europe. Immune to many illnesses (Plague, typhus)....Throve under persecution which would exterminate any other people.'[220] Memmi's research has emphasised the ambiguity of Semitic discourse regarding the spread of disease: The Jew was presented as 'an easy prey to certain maladies, slyly immune to others'.[221]

These eclectically rooted racial discourses about cowardice, conspiracy and disease had significant influence on the state and public reaction towards Jewish refugees fleeing Nazi Germany. Concerns about Jewish racial type ensured that whilst there was some sympathy for refugee Jews, especially as Nazi persecution increased, it was checked by the belief that Jews were partly to blame for the persecution which they attracted.[222]

Like most scientists, British society in general was appalled by the excesses of Nazi anti-Semitic policy. Even those scholars whose work has highlighted the uncertainty and ambivalence which dominated British reactions to Jewish refugees in the 1930s have observed the near universal British condemnation of Nazi anti-Semitism.[223] As Nazi aggression became more virulent, this condemnation increased further. Reactions to the 'Kristallnacht' pogrom in November 1938, which triggered a partial relaxation of immigration restrictions until the beginning of the war, have been described by Tony Kushner as a 'horror' which was 'spontaneous and deeply felt'.[224]

Despite this level of sympathy, there remained a widespread conviction in some governmental and parliamentary circles that Jews were guilty of many of the charges that were levelled against them by Germany and other countries, an idea that was at least partially rooted in perceptions of Jewish behavioural racial type.[225] As ideas about Jewish type tempered sympathy, it was only Hitler's excessively violent response which was deemed worthy of criticism. For example, Neville Chamberlain responded in 1939 to a journalist's question regarding alleged Nazi atrocities, that 'he was surprised that such an experienced journalist was susceptible to Jewish/Communist propaganda'.[226] To Chamberlain's mind, German Jews were to some extent playing up their plight, manipulatively trying to exact sympathy from Britain. A belief in such Jewish conspiracy permeated much of Britain's intelligentsia in the run-up to the war, and was nowhere more prevalent than amongst appeasers of Germany who increasingly viewed Jewish refugees as standing between British and German friendship.[227]

Whilst it would be an exaggeration to conclude that the entire British establishment was influenced by racialised images of Jewish conspiracy, these ideas were prevalent if not dominant within significant sections of British politics and media.[228] As one analyst has noted:

> British stereotypes of Jews were significant in marking them out as members of a group that was difficult, even dangerous, to help. Such prejudices helped to cast the image of the Jewish refugee in a problematic mould and thus to strengthen support for policies of restriction.[229]

The British establishment generally failed to perceive that Jewish people were simply innocent victims of false racial doctrine. Instead, racial characterisations of the community were invoked as explanations and mitigation for the violent treatment meted out to them, even as this treatment was criticised.

Whilst opposition to German ideas of race did sometimes lead to support for Jews fleeing Nazism, generally the effect of seeing the refugees through a racial lens ensured that Britain mostly excluded Europe's Jews. Racial discourses were so entrenched that even when faced with the humanitarian imperatives brought about by Nazi atrocity, the British establishment could not shake off ingrained racial images of Jews. As European Jewry floundered in the horror of the Holocaust, attitudes towards Jews altered little.

London has cited the response of senior Foreign Office civil servant Maurice Hankey to a programme proposed by the United States for the transmission of foreign exchange to save 70,000 Romanian Jews from Hitler's 'Final Solution' in 1943. Hankey could not escape from the idea that Jews were in some way attempting to profit from the international situation and were not as vulnerable as they seemed: 'I suspect the real object of the scheme is financial – Jews in Europe getting into dollars while there is still time.'[230]

Hankey's views, whilst not shared by everyone, were not isolated. In a 1950 report on 'alien immigration down to 1938', Home Office civil servant A.J. Eagleston argued that Jews had unnecessarily burdened the British Consulate in Vienna for visa appointments, not because they desperately needed to leave Austria after the Anschluss, but so they could sell any given appointment 'by auction to the highest bidder'. In a similarly racist reading of Jewish priorities, the author also recorded that Jewish capitalists in Germany sometimes did not come to Britain for refuge from persecution but instead 'to get behind the British tariff wall or to escape the difficulties which German foreign exchange regulations put in the way of export trade'.[231]

Eagleston's report also revealed that the Home Office racially assessed threatened European Jewish communities and ascribed to them different levels of suitability for immigration. In this way, Germany's Jews were generally regarded as better potential immigrants because they were 'thoroughly Westernised'. Jews from Austria and Eastern and Southern Europe were less suitable for life in Britain because it was perceived that they were 'much nearer to the Ghetto mentality'.[232] This idea of 'Ghetto mentality' was strongly rooted in the Semitic discourses which have been discussed here. For example, Jews were constructed by Eagleston as cowards who 'had no desire to make any kind of sacrifice, even for the noblest of causes, if they could avoid it'. Ultimately, Eagleston recorded, Jewish refugees were mostly 'shown by their character or history to be extremely undesirable and even dangerous residents'.

The idea of the 'harmful' refugee was partially rooted in stereotypes about Jewish behaviour but also in the idea that most of Europe's Jews were physically degenerated and diseased. There is evidence that the government was concerned that the infiltration of inferior Jewish types might pose a threat to the racial fibre of Britishness. As immigration policy evolved in the interwar period it became increasingly focused on the idea that Britain should gain from any immigrant admittance. Following (if unwittingly) the principles of Moul and Pearson, the government tried consistently to admit people that it thought would add to the stock of the nation. Whilst the moral importance of offering refuge to other less desirable 'aliens' was not ignored, especially in the 11 months which separated 'Kristallnacht' and the beginning of the war, these refugees were only allowed to enter on temporary visas on the clear understanding that their status was as 'trans-migrants', who would move on to final destinations as soon as this became possible.[233]

In this way, government policy towards 'aliens' can to some extent be described as eugenically informed or at least as being consonant with eugenicists' agendas. In his report, Eagleston noted that the admittance of refugees before the Anschluss was rooted in a policy of 'selection (in respect of both quantity and quality)'. Care was taken only to admit those who could offer something to Britain, and not those who it was perceived would take more than they could give. In this way, Whitehall aimed to ensure that Britain took only 'the cream of the refugees'.[234] This policy was, to Eagleston's mind, the evolutionary product of an immigration agenda which had been growing since the passing of the Aliens Act in 1905. As the policy evolved, not admitting 'aliens' except where this was 'to the advantage of the country' became the guiding principle.[235]

The 'advantage' that 'aliens' could bring to Britain was sometimes understood in economic terms. Immigrants were more likely to be welcomed if they had the potential to create wealth and jobs. London has pointed out that half the immigrants that were given leave to remain came from the professional classes, a fact which Eagleston's analysis shows to have been the result of a consistent strategy.[236] However, sometimes the idea of 'advantage' was rooted in more mysterious understandings of racial impact. For example, it seems to have been perceived that the harmfulness of the Jew to the British nation could be gauged through an assessment of the immigrant's level of distinct 'Jewishness'. As we have seen in the Home Office's favouring of German over other European Jews, there was a current of opinion which held that if Jewish qualities could be significantly removed from the immigrant (or were not prominent in the first place) then their presence in Britain could be racially advantageous or at least less onerous.[237] This idea may go some way to explain the government's decision in the wake of 'Kristallnacht' to allow the entry of 10,000 Jewish children (but not their parents) in what became known as the *Kindertransport* policy. For whilst these children may have still been perceived as racially different, there was time, given a different environment, to transform them into good Britons. London has argued in this context: 'To admit children was regarded as less onerous....The children would be Anglicised, growing up speaking English and thus less likely than adults to arouse xenophobia.'[238]

That the debate about refugee entrance was at least partly based in perceptions of racial desirability can be seen in the language of some of the social and scientific campaigning which took place on behalf of refugee Jews. For example, leading progressive scientists sought to help Jewish refugees by presenting them as racially desirable additions to British society, a defence reminiscent of Davies and Hughes in the 1920s. J.B.S. Haldane wrote in one 1934 paper that Germany had lost her scientific prominence since her removal of Jewish scientists.[239] In another article that year, Haldane made the further observation that Germany's loss could be Britain's gain. The issue, as he presented it, was very much a racial one:

> There is of course a strong case against the admission of persons of whatever race who are physically or mentally below the average. On the other hand the opportunity has arisen, as the result of recent political disturbances in Europe of admitting to British citizenship exiles of proved intellectual ability. Every eugenist (*sic*) should be prepared to recommend the admission to British citizenship of such exiles, provided that they attain a sufficiently high standard.[240]

Similarly, the popular novelist Louis Golding attempted in *The Jewish Problem* in 1938 to sell the Jewish refugees as a racial asset to Britain. 'The addition to any country of a body of young, strong, active, industrious immigrants with the probability of a long life before them (as the refugees from Germany and Austria in the main are) is clearly an asset, particularly to a land which, like England, is faced with an imminent fall in population.'[241]

Ultimately, it is not the intention of this chapter to suggest that scientific views on race were responsible for setting the public and state stance towards refugees. What is suggested in the case study of the Jewish refugees is evidence of Smith's 'common context' between science and society. In general, the public mirrored scientific distaste for the anti-Semitic violence of Nazi Germany. However, just as this distaste did not lead to the dismissal of race in science, neither did it lead most of the public and state to transform their views about the essential difference of Jews.

The state's refusal to provide refuge for the great majority of the Jewish refugees who wanted to come to Britain may have been mostly rooted in beliefs about Jewish people which did not come from science. However, the idea of inherent racial difference which seems to have influenced government refugee policy mirrored the dominant scientific perspective of the period. In the main, both British scientists and British society came and left the issue of race in this period with the belief that despite the wrongs of Nazi Germany, racial differences existed and mattered. It is this mirroring of core racial values that best describes the interaction between science and society in the 1930s. This was indeed a period where science was becoming more popular and where scientists increasingly could and did assume roles as leaders of public opinion. However, these new leaders were products of the society which had created them, and despite ostensibly bold projects like *We Europeans* they did not provide radical leadership on the issue of race. The war, however, would rejig this dynamic and bring forth a new level of scientific commitment and racial rethinking.

3
The Challenge of War: the 1940s

Scientists and the wartime racial agenda

The beginning of the Second World War brought forth a new level of social commitment from many of Britain's leading biologists. As we have seen in the previous chapter, the majority of scientific leaders were deeply opposed to the Hitler regime and the onset of war turned their opposition into action.[1] For some, like C.P. Blacker, this meant joining the military effort directly, whilst the older statesmen of British science found other ways to assist the Allied cause.[2] Julian Huxley, 52 years old in 1939, spent most of the first years of the conflict working unofficially as a British diplomat in the USA whilst Lancelot Hogben served the country by working on British Army medical statistics with Frank Crew for the War Office.[3] J.B.S. Haldane, some five years older than Huxley and Hogben, conducted pioneering research into human survival under water, designed to assist Britain's submarine effort.[4]

This commitment was fuelled and sometimes undermined by a fear that the war could end Western civilisation for generations. Of course, this fear was not exclusive to scientists, although their intimate knowledge of the potentials of the available weaponry must have made the prospect of all out European conflict extremely frightening.[5] Hogben, in particular, seems to have been afflicted by the fear that the war would destroy Europe once and for all. Ostensibly in America to conduct research at the beginning of the war, he was extremely reluctant to return to his post at Aberdeen. Hogben claimed that this was due to the superior American intellectual atmosphere and the higher value placed upon his work in the States, but his correspondence with Huxley reveals deeper fears:

> ...the world future seems so irredeemably black that it doesn't matter much what happens to any of us...I am quite willing to resign the

chance of a few months more useful scientific work and return to die with the other rats in the sewer, but I can't do it without money.[6]

Driven by ideology and fear, leading scientists tried to play a substantial role in the war effort, arguing that they could provide a unique service to the country.[7] As Haldane put it: 'At the present time we are fighting Hitlerism with bombs and depth charges. We should be doing so in the realm of ideas also.'[8] In one 1940 article, Huxley criticised both the government and fellow scientists for not realising how useful scientific expertise could be in defeating Hitler. He complained that 'many scientific workers have not been used at all, and others, who have for a long time been in Government service, or who have been recruited since the War, are not being used to their fullest capacities'. Huxley contended that underuse was partly the fault of government and partly due to 'the failure of men of science to realize both the power that lies in their hands and their responsibility for using it'.[9] In a 1941 article in the *New Republic* Huxley dismissed the possibility that a separate scientific world could operate apart from society during the war. 'The national emergency is so overriding that "background" science has to be given up, and the purest of scientists find themselves attacking the most practical of problems.'[10] Hogben also saw the situation between science and society in these terms and believed that scientists needed urgently to take sides, if science itself was to be preserved.[11] Writing before the war had started he prophesied:

> The time will soon come when scientific workers will be forced to choose between two alternatives. One is the social programme of the Fascist States, where pseudo-scientific rationalizations are advanced to withhold social privileges, restrict production, and so deprive science of the stimulus which it derives from expanding industry. The other is the extension of social privileges, the expansion of industry by increasing consumption, and the encouragement of science.[12]

With this desire to tie science and society together, those scientists who had led the progressive challenge to race in the 1930s aimed with new vigour to bring science to the public during the war. Haldane's assumption of the chair of the editorial board of the *Daily Worker* in 1940 provides a good example of the increasing crossover between leading scientists and society, even if the temporary government closure of the newspaper in January 1941 indicates that Haldane's belief in communist anti-Nazi struggle was not on message.[13] Above and beyond the shutdown of the *Worker*, Haldane's popular scientific writing in this period made him famous. Werskey has described him as 'a great public success' and has argued that 'as a stylist he had no peers among the popularizers of science'.[14] Similarly, Huxley's regular appearances on the *Brains Trust* radio programme made him a household name.[15]

The *Brains Trust* epitomised the new wartime closeness between experts and the general public. First going to air under the soon aborted title *Any Questions?*, it began as a radio show in January 1941, acquiring its lasting name after being dubbed as the *Brains Trust* by national newspapers.[16] It featured Huxley, along with the philosopher Cyril Joad from Birkbeck College and retired Navy reserve commander A.B. Campbell, although all three contributors dipped in and out of the show to allow other guest experts onto the panel. The programme followed a simple structure in which Joad, Huxley and Campbell answered random questions sent in by the general public. The *Brains Trust* ran for eight years and achieved a wartime peak of ten million listeners, second in public affections only to the Nine O'Clock news.[17] As an article in the *Listener* in 1960 put it, 'Tea with Joad, Huxley and Campbell became part of the life of Britain at war.'[18] As well as achieving massive ratings, the *Brains Trust* as a format spilled out across the country, as people staged their own panel question and answer sessions at local level. Sian Nicholas has described the programme as 'one of the most remarkable popular hits the BBC ever had'.[19]

The significance of the *Brains Trust* to this study is twofold. Firstly it propelled Huxley to an entirely new level of fame, giving him (and his views on race) a new social prominence.[20] Secondly, and perhaps of greater importance, the success of the *Brains Trust* highlights the desire of experts like Huxley and Joad to educate the public and a corresponding public willingness to interact with specialist knowledge. This combination gave science heightened authority in wartime Britain and ensured that those scientists who wished to write and talk about race could command an ever larger audience.

As had been the case before the war, Britain's leading biologists were determined in their wartime writing to attack Nazi racial theory. The first example of this commitment can be seen in the proceedings of an international scientific congress held in Edinburgh in the week before the war started. This congress brought together 600 biologists from 55 different countries, gathered to 'discuss the problems affecting the study of heredity and variation in many forms of life, including mankind'.[21] It resulted in the creation of a 'Geneticists' Manifesto', designed in response to the question: 'How could the world's population be improved most effectively genetically?'[22] The Manifesto was signed by nearly all of Britain's leading biologists. Whilst Gates's signature was notably absent, Crew, Haldane, Hogben, Huxley and the biochemist Joseph Needham all gave their names to the document.

The Manifesto considered other subjects as well as race, but devoted much of its attention to the issue. With a typically progressive focus on anti-Nazism and social reform it blamed the contemporary perversion of race on long-standing political inequalities:

> The second major hindrance to genetic improvement lies in the economic and political conditions which foster antagonism between different peoples, nations and 'races'. The removal of race prejudices and of the unscientific doctrine that good or bad genes are the monopoly of particular peoples or of persons with features of a given kind will not be possible, however, before the conditions which make for war and economic exploitation have been eliminated.[23]

Similar to *We Europeans*, the Manifesto did not attempt to destroy the idea of race but instead clearly separated Nazi racial theory from other uses of the concept. Admittedly the word itself was gone, only used in parentheses and with reference to the idea of race prejudice. Instead of racial characteristics the document described 'intrinsic (genetic) characteristics'.[24] Instead of races the Manifesto spoke of 'peoples'. However, in a trend that began with *We Europeans* and continued well into the postwar period, the end of the word did not signal the end of the concept. In fact, the Geneticists' Manifesto went on to outline at length the importance of 'intrinsic (genetic)' differences in divergent human populations, an analysis which added up to an ongoing belief in race only under another name. The key word in understanding this agenda in the Manifesto was 'monopoly'. The geneticists argued, as noted above, that no 'good or bad genes' were the 'monopoly' of any given 'peoples'. They did not argue that certain 'peoples' were not on the whole more advanced or more blessed with certain inherent characteristics than others, only that no one people had a monopoly on any particular gene. Thus, the Manifesto should not be read as an attempt to undermine the concept of racial mental difference. In the same way as Huxley had done in his 1936 Galton Lecture, the Manifesto merely emphasised the interplay of nature and nurture in character development, calling for 'recognition of the truth that both environment and heredity constitute dominating and inescapable complementary factors in human well-being'.[25]

In this context it seems arguable that Barkan has overstated his case by labelling the Manifesto as one 'in a number of anti-racist declarations'.[26]

It could in fact be argued that the rationale behind the Manifesto had little to do with a desire for race reform in itself and was more focused on the development of an international leftist scientific fraternity. This fits with Werskey's analysis of scientific engagement in the war, which he has described as being 'crafted with a view to consolidating gains that left wing scientists had already made in scientific and labour circles'.[27] The Manifesto after all blamed Nazism on 'economic and political conditions' and argued that wealth and opportunity disparity was responsible for preventing scientific assessment of hereditary difference:

> ...there can be no valid basis for estimating and comparing the intrinsic worth of different individuals without economic and social conditions which provide approximately equal opportunities for all members of society instead of stratifying them from birth into classes with widely different privileges.[28]

Far from being focused primarily on race, this language and agenda has been explained by Jones as an attempt 'to try and persuade' the Soviet delegation to attend the conference (which they did not do).[29]

Ultimately, the Manifesto called for international cooperation, a 'federation of the whole world, based on the common interests of all its peoples'.[30] This agenda resonated with the socialism of Haldane, Hogben and others, mirrored the kind of international liberal humanism which Huxley wanted for the future (and which would drive him to become the first Director General of UNESCO in 1945) and had little to do with any radical intentions concerning the idea of race.[31] As such, it set the tone of British biologists' contributions to the war. The Manifesto did not signal any innovative or definitive statement on racial difference, seemingly because the scientists involved felt that the quasi-radicalism of *We Europeans* and the like had gone far enough on this subject. It instead beckoned in a series of polemical wartime studies attacking Nazi racial theory, which collectively pushed British biology to a slightly more radical position on race. By subtly stopping its critique of the concept well before asserting that all peoples were equally endowed by heredity, the Geneticists' Manifesto epitomised the nature of biological challenges to race both during and after the war.

As was the case in the Manifesto, progressive biological writing on race during the war in general did not significantly move on the racial analysis of *We Europeans*. Wartime publications on the subject frequently referenced the earlier work and mostly endorsed its key arguments concerning the use of the term 'race', racial purity and Nazi racial theory. For example, in his 1941 pamphlet on 'The Nazi attack on international science', the progressive biochemist Joseph Needham asserted that further scientific elaboration on racial theory was unnecessary in the wake of *We Europeans*, which to his mind had brought a close to the subject. He argued that it was hardly 'necessary to devote any space to a consideration of the scientific refutation of racialism, since this has been very well done in books such as *We Europeans*, by Huxley, Haddon, and Carr-Saunders'.[32]

The race writing of most biologists in this period shared *We Europeans's* primary goal of discrediting Nazi ideology concerning Jews and racial purity. Leading progressives also forcefully echoed the *We Europeans* argument that allowing the continued use of the term 'race' carried a political and moral danger epitomised in Nazism. As Huxley put it in a letter to the *New York Times* in 1940: 'To continue to use the term "race"' was 'to play into the hands of the Nazis.'[33] In the final year of the war J.B.S. Haldane wrote a paper using the case study of Nazi racism as a lesson to humanity. Haldane warned his readers:

> So long as men and women believe in the superiority of their own race in anything like the way in which that belief was held in Germany, there will be a danger of aggressive wars and oppressions of minorities such as those carried out by the National Socialist Party of Germany.[34]

However, as in *We Europeans*, the progressive attack on Nazism did not equal a refutation of race as a scientific concept. A closer look at Huxley's *New York Times* letter reveals the ongoing ambivalence about the idea of race which continued to dominate progressive writing on the subject. Huxley argued in the letter, in language which was strongly reminiscent of *We Europeans*: 'The meaning of "race" has by now become so confused and distorted that the only satisfactory solution is to drop the term altogether.'[35] The key words in this sentence and argument are 'by now'. The point Huxley was making was not that race was never a biologically reasonable field of study but only that it was too problematic to use in the contemporary political context. Huxley advocated the dropping of 'the term'. This still did not equate to a willingness to drop the idea. In most wartime progressive writing the point being made was that the Nazis had oversimplified race and got it wrong, not that the concept or principle was erroneous or not worthy of further investigation. Haldane concluded in *Science in Everyday Life*: 'The truth about human races, when we know it, will no doubt be complicated.'[36]

Huxley and Haldane both retained strong beliefs in the existence of physical and psychological racial differences throughout the war, despite their leadership roles in the Allied anti-racial challenge to Hitler. Some of these racial views were ostensibly uncontroversial. When Haldane argued that 'we can certainly say that Englishmen and West African Negroes are different races, in the sense that you can always tell an Englishman from a Negro', he was making a case that (aside from the matter of his continued use of the term 'race' and his assumption that a black man could not be an Englishman) would not invite criticism from any quarter, either then or now.[37] However, both men retained opinions about physical racial types that went far further than Haldane's assertion that there were basic bodily differences between far-flung peoples. For example, both men continued to believe that there were similarly identifiable differences within European populations, though they argued that these did not correlate perfectly with any national population.

In response to a letter from an F. Twyman, who frequently corresponded with Huxley on the subject of physical types, Huxley confirmed his belief in the prevalence of certain head shapes in Germany. 'Of course there is a large component of what you call "flat heads" in Germany. These are the so-called Alpines.' Whilst he dismissed Twyman's further contention that 'Flat Heads' behaved in a certain way and that they were predominantly German, Huxley's engagement with arguments of this sort betrays his continuing belief in European physical racial types.[38] In another letter from 1941, Huxley dismissed the idea that there were such things as Aryan, Slav or Latin races (noting that these were 'language groups with no racial unity'), but conceded that both Mediterranean and Nordic races existed and that modern Italians were largely of a 'Mediterranean type'.[39]

Haldane's beliefs in physical types were similar. Writing immediately after the war, he remained prepared to contend, like Huxley, that it was 'quite reasonable to talk of a Nordic type', although he noted that it would be impossible to find a geographical area, aside from 'a number of Swedish villages' that was inhabited by 'pure Nordics'.[40] As late as 1950, Haldane was prepared to assert on record that Britons (along with Norwegians) inherited racial stock which predisposed them towards sea-faring: '…we haven't changed much racially. And we keep many of the good qualities of our ancestors. For example, both English and Norwegians make very fine sailors.'[41]

For both Haldane and Huxley, manifest physical differences between races served as an indicator that there were also likely to be corresponding temperamental differences. Huxley clearly outlined this perceived link between the physical and the psychological in *The Uniqueness of Man* in 1941:

> Ethnic groups obviously differ in regard to mean values, and also the range and type of variability, of physical characters such as stature, skin-colour, head – and nose form, etc.: and these differences are obviously in the main genetic. There is every reason to believe that they will also be proved to differ genetically in intellectual and emotional characters, both quantitatively and qualitatively.[42]

Haldane concurred in *Keeping Cool* that even though mental racial difference was not proved, 'it would be unscientific to say that there can be no connection between racial origins and mental characteristics in a country like England'.[43]

In the cases of both Huxley and Haldane, this belief in mental racial difference was accompanied by a corresponding belief in black inferiority. Whilst both mitigated this idea with the argument that 'overlapping' occurred between all races, nonetheless the average 'Negro' was still seen to be inferior to his European equivalent. Haldane wrote in his notes: 'Maybe on the whole English [are] cleverer or more moral than Negroes. Quite certain cleverest Negro cleverer [than] stupidest Englishman.'[44] Huxley concurred: 'I regard it as wholly probable that true Negroes have a slightly lower average intelligence than the whites or yellows.'[45] Occasionally Haldane expressed more extreme views on black inferiority. For example, in *Keeping Cool*, he dismissed the idea that certain black groups would ever achieve European standards of intelligence. 'The Australian Blacks are hunters. They huddle round fires in cold weather. But they had never thought of making clothes from the skins of the animals they killed. I find it hard to believe that their descendants will produce a Watt or Edison.'[46]

Ultimately, these leading progressives still believed in race and racial difference, both physical and psychological. No one could doubt their commitment to the fight against Nazi racism. In time, both men also went on to engage in similar battles against the racist policies of American segregation and South African apartheid. But Huxley and Haldane's political inclinations did not change their core scientific beliefs in racial difference. In their mind frame, it was vital to fight to ensure that race as a concept was not abused by illiberal regimes, but it was similarly important to prevent the abandonment of the concept and criticise the erroneous belief that all races were equal. Thus, in the wake of Nazism's perversion of race, Haldane was still prepared to write in defence of the concept:

> Some people are so impressed...with the harm done by Hitler's theory of superior and inferior races, that they think there are no inborn differences between races and individuals. Any baby could do anything if it was brought up in the right environment. I believe this is nonsense too.[47]

Haldane and Huxley may have been Britain's most famous progressive natural scientists in the war period but they did not dictate entirely the parameters of the anti-racial scientific challenge, even if their moderation and uncertainty did generally reflect it. The heart of the progressive case was still more rooted in opposition to Nazism than it was in post-racial analysis, but it is nonetheless true that some stronger anti-racial statements were made by progressive natural scientists in this period. For example, a more radical position on race is discernible in the writing of Lancelot Hogben (who now described himself as a social biologist) and Joseph Needham.

In Hogben's *Dangerous Thoughts* he challenged the idea that there were firm biological grounds for assuming that physical racial disparities were matched by psychological differences. Asserting that biologists were in no position to make any such argument, Hogben wrote that 'all existing and genuine scientific knowledge about the way in which the physical characteristics of human communities are related to their cultural capabilities can be written out on the back of a postage stamp'.[48] The key hallmarks of scientific post-racial theory are evident in this analysis. Hogben called for the dismissal of the idea of inherent psychological differences in recognition that race was, if anything at all tangible, only skin deep. In *Dangerous Thoughts*, physical differences between peoples were presented as being just that, not as signposts towards mental potential. Making a case for black and white equality, Hogben attacked the logic of scientific prejudice against 'the presence and absence of melanin in the deeper layers of the skin', arguing that to assess mental potential based on such information was as ludicrous as making a psychological judgement based on 'prejudice against freckles'.[49]

The corollary of this case was that differences in aptitude and behaviour between different human groups were the product of cultural environment. Whilst Huxley and Haldane cited the importance of the interplay between environmental and hereditary factors in shaping group behaviour, Needham and Hogben were prepared to champion environment as *the* key factor. In Needham's pamphlet on *The Nazi Attack on International Science* he conceded that scientists did not 'know enough' to decide for sure about the existence and causes of group intelligence differences. However, Needham's own take on difference was seemingly driven by the view that environment was all-important. He argued that present research indicated that the levelling of opportunity softened racial differences.

> In those parts of the world, indeed, where races meet in conditions of almost absolute educational equality, such as Trinidad, experienced teachers will freely admit that there are no detectable differences between the performance of Whites, Chinese, Indians, Negroes, and Caribs.[50]

In this analysis, there were no grounds for thinking that any person was 'more desirable than any other'.[51] Although for the most part, Needham's and Hogben's work on race read like Huxley's and Haldane's, this step further was important. It was, if nothing else, a natural scientists' confession that they did not have all the answers to the questions of racial difference that had become so important in this period. The progressives (or at least radical voices among them like Hogben and Needham) were beginning to realise that they, with their core beliefs in genetics, were not able to drive anti-racial science to its next level. This would require a reworking of the concept of mental racial difference as an entirely cultural (not a genetic) phenomenon, a transition that would in the post-war period mostly marginalise natural scientists and elevate cultural and social anthropologists and sociologists. Whilst genetics had, in the previous generation, played a key part in undermining earlier physical anthropological racial typologies, by the late 1940s geneticists had joined physical anthropologists in being perceived by social scientists as racially old fashioned. Their general lack of willingness to dismiss the biological role of race in shaping character and intelligence stood in sharp contrast to new social scientific theories which, with greater clarity, explained racial mental differences in social terms of culture and environment.

This transition of racial radicalism from genetics back to newer anthropological racial studies and sociology was driven by high-profile social scientific research on race in the USA. Here, the research of Franz Boas had spearheaded a new approach to anthropology, where the link between race and cultural behaviour was mostly undermined.[52] Across social scientific disciplines, a body of influential American scholarship asserted that the idea of mental racial difference was socially constructed and required a cultural (not a biological) explanation. In particular, *An American Dilemma*, Gunnar Myrdal's epic study of US race relations, did much to put this message at the centre of American thinking on racial difference.[53] Myrdal was a Swedish economist who was commissioned by the Carnegie Institute in 1938 in the hope that he, as a foreigner, could write an unbiased report on black–white relations in the USA.[54] Myrdal published his findings as *An American Dilemma: the Negro Problem and Modern Democracy* in 1944.[55] This book, described by King as 'the crowning achievement of modern sociology on an epic scale', attacked the idea of innate racial difference and argued that America's ongoing discomfort with the idea of white–black equality was at odds with her fundamental national values.[56] In doing so, it became perhaps the most valued point of reference in the battle against Southern segregation, cited by Chief Justice Earl Warren in the Brown versus Tapika case of 1954 and heavily influential on President Truman's Commission on Civil Rights in 1947.[57]

Importantly, Myrdal's work reflected the growing authority of American social scientists on the issue of racial difference. Other American studies of race in this period, notably the psychologist Otto Klineberg's *Characteristics of the American Negro* and anthropologist Ruth Benedict's *Race, Science and Politics* were similarly influential and contributed to an atmosphere where race, as an idea, was the subject of heightened challenge and analysis from outside biology.[58] In these social scientific studies, race was not set up as a physical reality but was instead largely analysed as a habit of social behaviour, built on economic, psychological and social foundations. Across the Atlantic, these new approaches were not going unnoticed amongst Britain's progressive natural scientists.

As early as 1941, Needham had argued that mankind was not suitable for analysis in the same way as other animals; that the complex social and emotional relationships of humans put them onto a different plane, in some ways abstracting them from biological assessment and discussion. In his words, 'man in his society constitutes a new and higher level of social evolution with its own laws and regularities'.[59] Needham considered that racial prejudice was a manifestation of this convoluted social interaction, not any inherent or biologically comprehensible tendency. On Nazi racial ideas of purity and difference he wrote: 'It is hard to avoid the conclusion that we have here a case for the psycho-analyst rather than for the teacher of elementary biology.'[60]

There is evidence that Huxley shared this creeping realisation that race could only actually be explained outside the field of biology. In his preface to Zollschan's *Racialism against Civilisation*, Huxley argued that cumulative cultural tradition constituted 'a second and much more rapid method of heredity and evolution alongside the normal biological one'.[61] In this period he wrote to Kenneth Little, Britain's leading anthropological specialist on Britain's black population (a scholar whose work straddled physical and social anthropology) asking for information about 'ethnic relativity'.[62] Little recommended reading to Huxley, notably the anthropology of Boas, and there is evidence that Huxley was at least partly persuaded by the argument. In a 1945 article he championed the thinking of Gunnar Myrdal, using it to explain the unimportance of race in a biological sense to an assessment of black achievement.

> Neither the failure nor the successes of the Negro in North America tell us anything about his possibilities in West Africa – except to remind us once more that human nature, so-called, is not unchangeable but can be developed or deformed by the conditions in which it has to exist; and that the structure and attitudes of whole societies are the overriding determinants of human fate.[63]

In the next chapter this study will argue that Britain's leading biologist progressives (along with more conservative geneticists and physical anthropologists) never went on to fully endorse new egalitarian anthropological/sociological positions on race and in fact became mostly opposed to them. The tentative reaching out to other disciplines that is evident here represents instead a final position for Britain's progressive biologists, one that did not alter in the 20 years after the Second World War. In recognising that biology did not have all the racial answers, Britain's progressives undermined Nazi racism and built a platform on which they would operate from this point onwards. Even for racial radicals like Hogben and Needham, an unequivocal refutation of the idea of group racial inheritance was a step too far. For Britain's progressives, making the argument that race was an idea *mostly* in the minds of people, corrupted and abused by many, was as much as they would do towards dismissing the concept. They wanted to help the nation beat Hitler. They did not want to destroy race totally as a biological category.

The Second World War greatly increased both the interest of Britain's biologists in the social and political world around them and their willingness to engage with it. It also led to a correspondingly amplified realisation amongst politicians that scientists could be very useful in the war effort. Perhaps most important in shaping the government's new warmth towards science was the substantial role played by scientific advisors in developing war strategy, tactics and weaponry.[64] Werskey cites large budget increases at the Department of Scientific and Industrial Research and the growth of the Parliamentary and Scientific Committee as good examples of political recognition of the role being played by scientists in the fight against Hitler.[65] Whilst some of Britain's leading biologists (like Haldane) worked well as military researchers and advisors, others played a different role as propagandists and diplomats for the Allied cause.

Joseph Needham and Julian Huxley both spent the first part of the war on unofficial propaganda tours of the USA. Huxley lectured on behalf of the 'Politics and Economic Planning' group, ostensibly discussing the post-war settlement, whilst Needham gave a lecture tour to US universities, before going on to China and eventually taking charge of a Sino-British science cooperation office.[66] In the States, both men were working for the British government, trying to persuade the Americans into the new war. This agenda, whilst secret, was not lost on Alex Carr Saunders, who wrote to C.P. Blacker concerning Huxley's planned trip: 'Julian is off to the USA to lecture – ostensibly – really to do a little indirect propaganda. All to the good – but it is pretty plain that the Americans are not coming in.'[67] The confidentiality of this cooperation between science and government was important because it enabled a quiet level of propaganda without generating the controversy of official emissaries, whose presence may have damaged American sensibilities at a time when many in the US were wary of being drawn into a new European conflict.[68] The fact that the British government felt that it could trust the likes of Huxley and Needham to represent the nation in this way, says something of the values and worldview that were shared between progressive scientists and their state in this period.

Scientific propaganda during the war most often took the form of written publications, which were sometimes conceived at the whim of specific scholars, but more often requested and paid for by the British government. One example of the direct political commissioning of scientific work was Julian Huxley's *Argument of Blood*. In April 1940, Huxley received a request from R.A. Bevan at the Ministry of Information. Bevan asked Huxley if he would be willing to write a pamphlet as part of a series that the Ministry was in the process of creating with Macmillan publishers. This series was designed, Bevan wrote to Huxley, 'to remind people in this country of the moral and spiritual values which we are defending, and to attach a really constructive meaning to the words Liberty and Democracy'.[69] It emerged as the Macmillan war pamphlet series, tackling a whole range of moral, political and scientific issues and including pamphlets from a raft of well-known intellectuals and scholars, including writers A.A. Milne, E.M. Forster and Huxley's *Brains Trust* colleague Cyril Joad.[70]

Huxley's brief was to write a political comparison of Allied and Nazi science, addressing 'the freedom of scientific research in this country compared with the restrictions imposed by the Nazi State'.[71] It is safe to assume that Huxley's selection for this task was not based upon his having any specific knowledge of this field but instead on the fame of his name. Whilst readers were advised at the beginning of the pamphlet that 'no one is more competent to expose this wholly unscientific conception than Julian Huxley', in private the Ministry recognised that Huxley perhaps did not have the information to write the pamphlet, reassuring him: 'You may rely on us to assist you to the best of our power in procuring any material you may require.'[72]

As with *We Europeans*, five years earlier, Huxley was politically sold on the project but did not have the time or maybe the inclination to actually do the work himself. He told the Ministry of Information that he wished he could have helped, but that he had 'not been at all well' and had 'a great many commitments'. However, in recognition that the real issue here was the use of his name, Huxley made a suggestion: 'If it were possible to get the actual writing mainly done by somebody else and my name were any good, I would be glad for it to be used', though he added that 'in general I do not think this sort of arrangement is satisfactory'.[73]

In the following month Charles Singer met with the novelist Graham Greene (who was seconded at this time to the Ministry of Information) and agreed to write the pamphlet on Huxley's behalf, under the title 'Argument of Blood'.[74] Given that Singer and Huxley were good friends and that Singer had played a similarly worthy but thankless role in the creation of *We Europeans*, it is unsurprising that Huxley chose to recruit him for this task. As had been the case in 1935, Singer shared Huxley's animosity towards the Nazi regime and was prepared to work (behind the scenes if necessary) to undermine it. In its published form, the work was everything that one might have expected from such a project. It was both emotive and dramatic, showing its clear propagandist intentions from the outset. *Argument of Blood* painted the Nazi regime as the destroyer of science:

> We have watched a great country, once justly proud of her scientific achievements, abandoning her belief in reason, pouring scorn on the disinterested pursuit of truth, expelling a quarter of her own distinguished scholars, persecuting the scholars of the territories she has invaded, rendering science servile, humiliating the profession of learning, and lowering the whole level of civilisation.[75]

Unsurprisingly, *Argument of Blood* mostly focused on the subject of race (as the title suggests). In language reminiscent of *We Europeans*, Nazi racial theory was dismissed as a 'pseudo-science... based on a series of propositions which no serious man of science can accept'. The pamphlet railed against Nazi teaching of racial biology, described as 'baseless and scientifically worthless', and ripped apart Nazi scientific racial output, most notably Lothar Tirala's *Race, Mind and Soul*.[76]

That such a study should use a critique of race to attack Germany is hardly surprising, given the political context of publication and the nature of its antecedents. The pamphlet seemingly reflected a governmental and scientific willingness, documented above, to oppose Nazi racism.[77] However, at a deeper level, *Argument of Blood* tells us much about the tensions that were prevalent among British scholars of race during the war and serves as an example of the 'thus far and no further' stance of Huxley and other leading biologists that has already been noted.

Initially it seemed that Huxley and Singer would have no conflicts over the material in the pamphlet. Huxley after all had been more than happy with Singer's work on *We Europeans*. However, when Huxley read through the pamphlet that was to carry his name, he was not satisfied with some of its content. In a letter to Singer, Huxley outlined his objections, which, he wrote, were shared by Graham Greene at the Ministry. Huxley criticised Singer's understating of pre-Nazi German scientific achievements, suggesting that these should be emphasised in order to 'bring the present bad situation into higher relief', and told Singer to write more about the German abuse 'of science and scientists in occupied territories'. More significantly, Huxley criticised the nature of Singer's attack on German racial studies.

> We also felt that instead of dismissing racial biology and their versions of anthropology as you do, as if they were not worth

serious consideration, there should be about a page on these, showing to what depths of distortion and folly you can get when official dogma interferes with free scientific teaching and learning.[78]

These criticisms reflected something of the nature and limitations of the racial challenge put forward by Huxley, other progressive biologists and the state in this period. The intellectual attack on Nazism in *Argument of Blood* was not intended to be an attack on race theory itself but on the Nazi perversion of it. As such, Huxley felt that Singer's writing should emphasise the Nazi destruction of previously valuable research and the attacks against those good scientists who had fallen under Nazi control. Having thought that he could trust his friend to parrot his own views over these issues, Huxley found that Singer's attack on race went further than he wanted to go. In the end, Huxley reworked much of the material, to the extent that he wrote to the Ministry of Information demanding payment. He complained to Greene:

> When I said that I would be willing to put my name to this pamphlet, I thought that Singer would be able to do the work quite adequately. As it turns out, I must have put in many hours hard work on it. I think that if there is any money available I should receive something. Could you let me know if this is possible?[79]

As for Singer, he seems to have been mildly annoyed by Huxley's quibbling but uninterested in the small intellectual distance between them. Singer felt that the most important thing was to get the work into the public domain where it could serve its intended function. He told Huxley: 'By all means alter it in any way you like. We can discuss points if you like but, better still get it out.'[80] In the wake of publication, neither scholar felt that the final work was his own.[81] However, years later, in marked contrast with Huxley's take on the work, Singer did claim *Argument of Blood* and underplayed any discord that had occurred:

> I wrote the pamphlet, but after discussion it was decided that it had better go out in Julian Huxley's name as being better propaganda. Nevertheless the article really is mine, and if Huxley altered it at all, which I cannot remember that he did, it was only just here and there in a turn of phrase.[82]

The relevance of the limited disagreement between Singer and Huxley on this project is twofold. Firstly, similar to the analysis above which compared the racial perspectives of Huxley and Haldane with Needham and Hogben, *Argument of Blood* reveals the limitations and confusions of the progressive natural scientists' attitude to race. In the same way that this study has argued that sociologists and anthropologists were ultimately needed to take anti-racial analysis to the next (post-racial) level, here the historian of medicine, Charles Singer, was cast as radical, wanting to go slightly further in dismissing race than Huxley or the Ministry of Information. Secondly, this text indicates the extent to which leading progressives like Huxley came together on race and science with the agenda of the British government. Although Huxley could not spare the time to write the text that Greene and the Ministry of Information wanted, in terms of message there was a clear and shared terrain of interest between the government and men like Huxley, who, after all, was working for them in this period.

This cooperation seems to indicate the kind of mirroring of views between leading scholars and the government which was presented above in the analysis of science and society in the interwar period. However, it is arguable that during the war, scientists did not merely mirror the dominant political perspective on race, but tried to move forward the agendas of the government and thereby the nation on some racial matters. This idea of scientific leadership can be explored more thoroughly by considering the stance of leading biologists towards two key aspects of wartime policy where issues of race seem to have been significantly involved. One of these issues concerned the treatment and internment of enemy 'aliens' in 1939 and 1940; the other, the wartime utilisation of black British subjects in Europe.

Scientists versus the state? Enemy aliens and Britain's black subjects

As Britain descended into war with Germany, there were increasing questions about what should be done with those enemy nationals that were living in the UK. Britain had responded to this issue in the First World War by interning and expelling what the state termed 'enemy aliens'.[83] This war experience formulated a precedent concerning how to handle 'enemy aliens' in Britain in the event of another conflict.[84] However, as the Second World War began, there was only limited pressure for a policy of internment. A general recognition existed that most 'enemy' nationals living in Britain were people who had fled the Hitler regime, mostly Jewish refugees, who were not likely to pose a threat to the state.⁸⁵ The Home Office, which retained primary responsibility for internment strategy, was not inclined to pursue any policy resembling general internment. However, local tribunals were set up across the country, charged with assessing the loyalties of German and Austrian 'aliens' and with registering them according to the risk they posed to British security.⁸⁶ 'Aliens' could be classified with the letters A, B or C. Category A was reserved for those with known sympathy for enemy regimes, whilst those in category C were deemed to pose little or no security risk. Of the aliens considered by the tribunals, 66,002 were classified in this bottom category, 6,782 in the intermediate category B and only 569 in category A.⁸⁷ Despite the classification of the great majority of 'aliens' in the lowest security category, confusion about the process and local prejudices seem to have led to the misclassification (and thus to the internment) of many 'friendly aliens'.⁸⁸

As the war continued, attitudes towards internment changed drastically. The dramatic Nazi invasions of Norway, Holland, Belgium and France created a fear that German success had been fuelled by 'fifth column' activities in invaded countries. This belief resonated through much of the press and many government agencies, creating a corresponding pressure to intern enemy nationals living in Britain.⁸⁹ Whilst the Home Office remained unsure, other agencies, notably the War Office and various security committees, demanded a policy of general internment.⁹⁰ As a result, between May and June 1940, over 25,000 'aliens' (of which the majority were Jewish refugees from Nazism) were placed in internment camps around the UK.⁹¹ Some attempts were made to transfer 'enemy aliens' to other Commonwealth countries.⁹² In July 1940, German torpedoes sank the *Arandora Star*, a ship carrying Italian and German internees bound for Canada, leading to the deaths of over 600 'alien' passengers.⁹³ In another high-profile incident, three crew members were court martialled after allegations of abuse and anti-Semitism on another ship (the *Dunera*) which took both Jewish refugees and Nazi sympathisers to be interned in Australia.⁹⁴ By the end of 1940, partly because of the stories concerning these ships and partly as the threat of imminent German invasion waned, most internees were released and more liberal government and press voices reasserted themselves.⁹⁵ Many 'aliens', however, remained in internment camps for much longer periods, as their security risk status was reconsidered.⁹⁶

Ostensibly, aliens were interned on the basis of their nationality and an objective criterion of risk, irrespective of their racial background. In this way, the interning of refugee Jews along with other Germans was presented as a strictly non-racial, if unfortunate, aspect of government war policy. It seems arguable that non-discerning Germanophobia and not Judaeophobia was the primary driving force behind the internment.[97] However, the truth is somewhat more cloudy. As Kushner has argued: 'Although the aliens were not interned because they were Jewish, neither was their Jewishness irrelevant.'[98]

Looking more closely at internment policy, two strands of racial reasoning seem visible within government decision making: the first strand held that there was a real risk that the 'Jewishness' of refugees would make them unreliable allies (and possibly even traitors) in the war struggle; the second that the presence of Jews (and the inevitable racial behaviour of these Jews) might increase British popular anti-Semitism to a dangerous level. Both these contentions were based at least partly in the racialised perceptions of Jewish character which were prevalent in much of the anti-Semitic discourse of this period. It is arguable that parts of the government acceded to the lure of these racial discourses and that at times there was a current of acceptance of racial Jewish images within government decision-making circles.

Many parliamentarians and government members seem to have been convinced that refugees in Britain posed a 'fifth column' risk to the nation, based on reports (like those of Neville Bland) concerning Nazi success in forcing the capitulation of European neighbours.[99] Earl Winterton told the House in August 1940: 'Again and again in the countries on the continent which were invaded by Germany it was found that refugees aided Nazis in their march.'[100] Bland's report, whilst not making references directly to Jewish refugees, implicitly (possibly deliberately) alluded to a specifically Jewish threat. His famous comment, that betrayal could come from 'the paltriest of servant maids' or 'every German or Austrian servant', was surely made in recognition that the overwhelming majority of 'aliens' in domestic service in Britain were refugee Jews, whose right to remain in the UK had been made conditional on their agreeing to take on such work.[101]

As Europe seemingly crumbled in the face of Nazi aggression (and bearing in mind that Britain was the likely next target of Hitler's expansionism) it is not surprising that fears of invasion and betrayal came to the fore. However, the specific manifestation of betrayal concerns indicate a widespread acceptance of racial stereotypes within the government imagination. Images of the Jew as a potential 'fifth columnist' seem to have obscured the real status of the refugees.[102] Colonel Wedgwood was for the most part isolated in the House of Commons internment debate of August 1940 when he pointed out: 'Nobody can doubt that of all those who are on our side, the Jews are most interested in our victory and are the least likely to act in any sort of way as agents of the enemy.'[103]

There does not seem to have been any substantial evidence to suggest that refugees had 'betrayed' their host countries in the face of Nazi attack. Again, Wedgwood was left to point out that only a racist caricature linked Jewish refugees to espionage allegations.

> I would very much doubt whether there has been any single case in this country of proved enemy assistance from any Jew. Indeed I know that there has not been, but there have been whispers.[104]

Comments from other parliamentarians seem to support Wedgwood's allegation that members were being influenced by racial images of Jews. Mavis Tate (MP for Frome) cited the overtly anti-Semitic writer Douglas Reed in making her case for the mass internment of refugees:

> I sympathise with the Jews, but Germany has learned to make skilful use of them. There is a book called 'Nemesis' by Douglas Reed. He pointed out, very clearly, in several books, what was going to happen, and the use made, in some instances, of Jews by the Nazis. It is no good saying that because a person is a refugee, because a man is a Jew and a victim of Nazi aggression, that he may not, nevertheless, be a potential danger to this country.[105]

Implicit within spying allegations lay the idea that Jewish immigrants (even when well meaning) were unlikely to have the courage required to withstand the terror of an imminent invasion.[106] The notion that Jews panicked easily and evaded any dangerous responsibility in wartime had been prevalent (as we have seen) in the First World War and soon found popular currency as the new war began.[107] These prejudices permeated decision making at the highest level during the war.[108] Refugee campaigner and MP Eleanor Rathbone highlighted the government tendency of thinking 'racially' in a leaflet designed to challenge anti-Semitism in policy making in 1943. State attitudes, Rathbone alleged, were formed against the stereotypes that 'the Jews are cowards; they shirk military service...the Jews are panicky'.[109] There seems to be some basis for Rathbone's allegation of governmental prejudice. After the German invasion of Czechoslovakia, one Foreign Office official communicated the idea that refugee Jews had not needed to leave the country and had only done so because of their lack of nerve. D.P. Reilly's notes for Lord Halifax on Jewish Czech refugees who had fled to Poland recorded:

> A great many of these are not in any sense political refugees who were in danger, but Jews who panicked unnecessarily and who need not have left. Many of them are quite unsuitable as emigrants and would be a very difficult problem if brought here.[110]

Utilising similarly rooted racial stereotypes of Jewish cowardice, the Central Office for Refugees wrote to the Board of Deputies in April 1940, explaining that internment was necessary to restrain 'people of weak moral character...who might succumb to temptation in time of war'.[111] Even the usually liberal John Anderson (in the House of Commons internment debate of August 1940) questioned whether family and interests left behind in Germany might lead refugees 'perhaps at the hour of our greatest peril, to take action, on an impulse it may be, which afterwards they might greatly regret?'[112] Foreign Office minister Peake likewise concluded in this debate: 'It is necessary to have people of faint heart out of the way.'[113] It was in this atmosphere that male enemy aliens were interned, albeit briefly, en masse.

British scientists, for the most part, took a different line from the government on the internment issue. From the beginnings of Nazi Germany's persecution of its Jewish population, many leading British biologists spoke out and tried to help those of their scientific colleagues who were suffering under the regime, a tendency which later spilled into attempts to defend internees. Haldane, for example, addressed the students' union of the University of London within months of Hitler's accession, attacking the anti-Semitism of the regime.[114] As the war approached, numerous British scientists worked to secure residency and employment for exiled Germans. Huxley's papers include grateful correspondence from Woburn House, the administrative centre of Anglo-Jewry, thanking him for feeding and housing refugees.[115] In 1940, he became patron of the Free German League of Culture's Committee for the Evacuation of Central European Refugee Children in Great Britain.[116] This was far from exceptional behaviour. Charles Singer was involved throughout the war with the Central Office for Refugees.[117] For Singer, the declaration of war, which terminated the arrival of significant numbers of refugees in Britain, signalled an opportunity to return to work, refugee assistance having become his main preoccupation. He wrote: '...the war has enabled me to work again. The stream of refugees has, of course, cleared and my job to do medical work when I am needed, is clear.'[118]

The efforts of individual scientists echoed those of scientific bodies such as the Society for the Protection of Science and Learning (SPSL), which had grown out of the Academic Assistance Council (AAC), formed with the goal of aiding 'university teachers...of whatever country, who, on grounds of religion, political opinion or "race" are unable to carry on their work in their own country'.[119] The AAC was founded as a result of discussions between William Beveridge (then the Director of the London School of Economics), the theologian James Parkes and Hungarian physicist Leo Szilard in 1933, all of whom felt that non-German institutions could benefit from Germany's exclusion of many of her top academics.[120] The work of the AAC was rooted in the Royal Society, under the leadership of A.V. Hill, who gave the AAC, along with his support, their first office space.[121] The efforts of the AAC/SPSL led to the placement of over 500 German academics.[122] These scholars were not only housed in British universities. With the help of Joseph Needham, the SPSL negotiated with the Rockefeller Foundation, enabling over 160 German academics to find new posts in the USA.[123]

The efforts of the SPSL on behalf of refugees became doubly significant after the onset of internment policy. The British government was in principle prepared to consider for release from internment camps those scholars who 'prior to internment were doing work of immediate national importance, the resumption of which would be of direct benefit to the country'.[124] To this end, the SPSL was told that the Home Office would in this matter take account of recommendations made by a tribunal of the Royal Society.[125] However, only 'two or three' out of over 500 scientists recommended in August 1940 were in fact released from internment following the Royal Society's intervention, indicating a distance between the SPSL/Royal Society and the government over the internment issue.[126] It seems here that scientists were working to drag the government into a more enlightened policy. A letter from the SPSL to Huxley in November 1940 reveals their perspective: 'Fortunately most of the scientists are now released from internment and we have every hope of getting the rest out by the end of the autumn, thanks to Professor A.V. Hill's unflagging efforts.'[127]

That there was a level of scientific disaffection with government policy on internment is evident in Huxley's work on behalf of refugees in this period. Huxley took up the cases of numerous scientists who he thought should be released from internment camps. For example, at the height of internment in June 1940 he wrote a series of letters on behalf of a German biologist Hans Honigmann. Huxley wrote to Lord Horder, F.A. Newsom at the Home Office and the Commander of the internment camp to which Honigmann had been sent, demanding his release.[128] Huxley was moved to act partly because Honigmann had been his research colleague four years earlier, but he also wrote numerous letters for other interned scientists in 1940, behaviour seemingly rooted in genuine irritation about the inadequacies of internment policy.

At one level, Huxley's problem with internment policy was rooted in humanitarian disgust. He complained that conditions in internment camps were 'scandalous and ought to be remedied' and lambasted the government for the callousness of providing the German authorities with the names of interned refugees, through the Swiss legations, as prisoners of war.[129] Looking back on internment policy in a 1941 article, Huxley concluded: 'Practically every thinking Briton feels shame at this decision and its results.'[130] Above and beyond the humanitarian issue of interning enemy nationals, Huxley felt that Britain had wasted an important resource by interning scientists. He told Honigmann's wife: 'I am very sorry to hear that your husband has been interned instead of the authorities making use of him in this country.'[131] Huxley wrote numerous articles calling for refugees to be utilised effectively in the war. At the height of the internment crisis, he wrote to the *Picture Post* arguing that refugees could and should play a crucial role in the fight against Nazism.[132] Huxley contended that the war was about ideology, and that as a result individual actions would not be based upon nationality or race. This was a war, he argued in 1941, not so much between sovereign states 'but a horizontal war between groups that quite transcend national boundaries'.[133]

Huxley's stance on internment and refugees reflected a wider scientific trend in the UK. Amongst professional groups, many of whom were almost openly hostile to refugees, scientists and academics in general were radical in terms of trying to assist and incorporate victims of Nazism.[134] It seems reasonable to argue that changing scientific thinking on race was important here. Whilst the government could not escape from being racially suspicious of the mostly Jewish refugees, feeling that their Jewishness posed an inherent risk to the UK, scientists like Huxley, Haldane, Needham and others were able at least partly to transcend such thinking. As a result, scientists did not in this instance mirror wider social views. Over internment, they actively tried to move the government to a more racially enlightened position, which they felt was not only humanitarian but also essential to maximising the war effort. The internment issue does not provide the only wartime example of this kind of leadership. Over the issue of black people in Britain, and the wartime handling of the Empire, some scientists similarly attempted to move the government to a more progressive policy. However, as we shall see, the already highlighted beliefs of the scientific progressives in black racial difference and inferiority tempered their leadership in this area so that it did not match the often bold attempts to free refugee Jews from the racial stigma that was attached to them.

Government thinking on how black people could be used in the war was largely underpinned by prevalent social beliefs about black racial difference and intellectual inferiority. It is clear that the government generally viewed any increased black presence on the British mainland as undesirable, mainly due to fears about the sexual danger posed by black men to white British women and the effects this could have on the racial make-up of the nation. Nonetheless, Britain's self-image as the centre of an Empire and wartime need combined to ensure that the presence of black volunteers, workers and troops would not be entirely avoided. For one thing, Britain could not afford to send such a negative message to the Empire, especially in the face of rising nationalism in Africa, India and the Caribbean. For another, the fight against Nazi Germany forced upon Britain the mantle of anti-racist moral authority, in opposition to the extreme racialism of her foe.[135] In the midst of these competing considerations a small number of black workers were allowed to enter the UK. It was beyond the powers of the British government to prevent the arrival of the tens of thousands of black US GIs that followed them.[136]

There is considerable evidence that government responses to the heightened black presence were affected by racial considerations. When the decision was taken to allow a small number of black Honduran men to serve as war workers on a forestry project it was felt necessary to house them in remote quarters (as far away from white women as possible). The director of the home-grown timber department said that he did not 'feel inclined to take responsibility for placing these men...on private estates close to houses and cottages occupied by estate employees'.[137] When the Duke of Buccleuch (on whose estate the men were working) wrote to the government asking about what could be done to prevent racial sexual mixing taking place, Harold Macmillan responded for the government that there would 'naturally be the risk of some undesirable results...All we can do is mitigate the evil.'[138] The Minister for Supply was even more candid. Although the government took no interest in sexual behaviour 'where Europeans were concerned', it could not be so passive 'where coloured people were involved'.[139] In 1942, the police raided the camps where the Hondurans lived and arrested all the white women present. Whilst there remained no legal barrier to interracial sexual contact, this incident was just one of many examples where coercive action was taken by state authorities against white women who consorted with black men.[140]

Some members of the government wished to go further and entrench anti-'miscegenation' law. In correspondence with Churchill, the minister for war Sir James Grigg outlined a firm desire for legislation of this kind. Far from just producing information that would discourage black/white mixing, Grigg told Churchill that there was a necessity 'for measures more stringent than education'.[141] In a private report for the Cabinet, the minister outlined his stance against 'miscegenation'. 'White women should not associate with coloured men. It follows then that they should not walk out, dance, or drink with them.'[142] Grigg justified this case with repeated assertions that black sexual activity with British women could have a disastrous effect on the morale of British soldiers serving abroad. 'I expect that the British soldier who fears for the safety or faithfulness of his women-folk at home would not feel so keenly as the BBC and the public at home appear to do in favour of a policy of no colour bar.'[143] Grigg's failure to win over his colleagues and secure legislation against 'miscegenation' should not be seen so much as an indication of governmental disagreement with the minister, but more as a reflection of the British reluctance to legislate in overtly racial terms.

Looking at newspaper coverage of alleged sexual contact between black troops and white women, one cannot escape the feeling that reports were often little more than racialised perceptions of exciting, promiscuous black eroticism. A report from the *Huddersfield Examiner* told its readers: 'The problem is one of white girls and coloured men meeting clandestinely and making love to one another in shop doorways, quiet side streets, open spaces and in some instances in vehicles drawn up at the side of the pavement.'[144] The sociologist Anthony Richmond, in post-war analysis concerning the difficulties of black people in wartime Britain, emphasised this problem of a racial super-eroticised black image.

> Among the stereotypes are those which attribute to the Negro an abnormally high sexual potency, a tendency towards promiscuity and a high capacity for giving and receiving sexual satisfaction from

intercourse. It is often suggested that a white woman who has once had intercourse with a Negro will never return to a white man.[145]

These fantasised perceptions seem to have affected thinking at every level of policy making. Marlborough wrote to Churchill that the danger of a black troop presence was their possession of marijuana that 'could excite their [women's] sexual desires either as a cigarette or ground up in food'.[146] Grigg told the Prime Minister that black sex crimes were inevitable due to 'the natural propensities of the coloured man'.[147]

Closely aligned to fears of sexual mixing was a corresponding racial belief that 'blackness' inferred the presence of diseases that could infect British society if any degree of intimacy were allowed. One reason frequently given for excluding black people from Britain was that they were carriers of venereal disease. It was this belief that probably lay behind Grigg's 1943 assertion that black troops were 'something of a commitment from the health point of view'.[148] When black workers were allowed to come to the UK, special VD clinics were often set up for them in an attempt to prevent the spread of the disease.[149] British Intelligence reports noted that female war volunteers (in the Women's Voluntary Service) were reluctant to help black soldiers because of the 'prevalence among them of venereal disease'.[150] Eventually, special separate 'silver birch' clubs were set up to entertain black soldiers in a safe 'controlled' environment.

Opinions about black racial inferiority seriously affected government policy during the war. In particular, the decision not to utilise black soldiers from the Empire in European combat (except in the most menial and peripheral ways) seems to have been rooted in ideas of this kind.[151] In 1942, the Minister for War prepared a report for the Cabinet concerning the possibility of utilising black military manpower. The report urged the Cabinet not to employ black soldiers in the European fighting arena because they were inferior to white troops. In a two-page endnote (marked 'not to be published'), Grigg offered a candid explanation of his opinion as to why black soldiers should not be allowed to fight.

> While there are many coloured men of high mentality and cultural distinction, the generality are of a simple mental outlook...In short they have not the white man's ability to think and act to a plan...Too much freedom, too wide associations with white men tend to make them lose their heads and have on occasions led to

civil strife. This occurred after the last war due to too free treatment and associations which they had experienced in France.¹⁵²

Grigg not only saw black troops as mentally inferior to their white counterparts but also as a physical liability. He wrote to Lord Stanley in 1943 that West Indian soldiers could not be used as combatant troops because they were 'not a very robust race'.¹⁵³

These views were not merely held by Grigg, but were common currency amongst Whitehall officials and other politicians.¹⁵⁴ Even war service that was considered less prestigious was often denied to black Britons amid allegations that they were not mentally up to the task. A Ministry of War minute sent to the Colonial Office reveals racial reasoning behind the rejection of Caribbean sailors from the 'Seamen's pool'. The note recorded the view that 'West Indian seamen are not as a rule up to the normal standards of skill and discipline [and that] during the war they have adopted a decidedly "cheeky" attitude'.¹⁵⁵ A 1943 War Office report entitled 'The employment of British West Indian soldiers in a theatre of war' concluded that these soldiers were undesirable combatants within every potential arena of conflict.¹⁵⁶ Black troops in the past had proved 'quite unsuitable' and 'their staying power under aerial bombardment [was considered] untried and suspect'. In West Africa, the troops would be 'nothing but a nuisance', in India 'an undoubted embarrassment'.¹⁵⁷ Ultimately, and tying in with discourses concerning black primitivism, the report concluded that the armed forces may 'lose control' of black troops, should they attempt to utilise them.¹⁵⁸

Given the presence of these kinds of beliefs it is not surprising that the plan to form a distinct West Indian Regiment repeatedly stalled. A note from the Colonial Office recorded the excuses used by the War Office for delaying its establishment: 'The War Office maintained their objections to a combatant unit...putting forward objections on the ground of climate, accommodation and shipping.'¹⁵⁹ Black people also found themselves quietly prevented from entering other active military roles by various sections of military regulations, which enabled the exclusion of people from non-European backgrounds.¹⁶⁰ However, in 1939, these official racial restrictions were finally removed. This did not, though, signal a new period of egalitarianism within the armed forces, but merely a shift from overt to covert exclusion. As Lees at the Colonial Office noted regarding the removal of these restrictions: 'This does not, of course, mean that British subjects who are obviously men of colour will in practice receive commissions.'¹⁶¹ Sherwood has demonstrated the widespread continuation of racial exclusion across all three armed services during the war and within British military support units.[162] She has cited the following 1941 advertisement for the Auxiliary Territorial Service (ATS), as offering rare, direct evidence of the continuing operation of such restrictions. Appearing in *The Scotsman* in 1941, the advert read: 'Women between the ages of 18 and 40 are invited to enlist for general service overseas in the ATS. They must be British subjects of pure European descent.'[163]

Amid prevalent racial perceptions of black inferiority, the prospect of black soldiers playing a significant role in European conflict was discounted. Although a limited number of Caribbean black technicians and other volunteers were allowed to enter the UK and whilst a 'West Indian Regiment' was eventually formed, overall the government had no desire to use any significant number of black Britons in European war service.[164]

The track record of Britain's leading progressive scientists reveals some willingness to defend the rights of black people in the face of political oppression. Huxley and Hogben both cut their teeth on white/black relations by travelling and working in Africa in the interwar period. Hogben's work in Cape Town between 1926 and 1930 brought him into contact with the kind of racial beliefs that would soon drive the apartheid movement.[165] He was extremely hostile towards these calls for segregation in South Africa and ran a household that was, his biographer recalls, 'openly anti Apartheid'.[166] It seems indeed that it was Hogben's disaffection with South African racial politics which led him to return to England, and a lectureship in Social Biology at the LSE, in 1930.

Hogben explained his leaving Africa in these terms, noting that it was an 'earthly paradise', which would have made for 'the happiest part of my professional career had it not been for the fact that a lately-elected Nationalist government was laying the foundations of Apartheid'.[167] He later penned vicious attacks on South African race policy, most notably in *Dangerous Thoughts* in 1939 where he mocked the 'chromatocracy' that existed there.[168] This contempt for the use of race in South African politics was to some extent radical but was shared by some of his colleagues in British biology. For example, J.B.S. Haldane responded to the claims of Sir Ernest Guest (a former minister of Southern Rhodesia) that white political dominance was justified by superior heredity by noting that this was 'a plain lie'. Haldane's writing often attacked the scientific logic of the 'colour bar'.[169] From his increasingly Marxist perspective, he perceived that the science of race was being manipulated by politicians across the world to justify exploitation, as part of an intellectually and morally unjustifiable strategy of capitalist exploitation spanning South Africa, the British Empire and Nazi Germany.[170]

Although he did not share Haldane's wider beliefs about the bankruptcy of capitalism, Julian Huxley was similarly unimpressed by the moral stance taken by white settlers towards their black countrymen when he toured Africa in the late 1920s. On returning, he published *Africa View*, which was far from egalitarian in its racial outlook but also repeatedly critical of white colonial attitudes. Attacking white exploitation of the continent, Huxley argued: 'You cannot expect a people to make a really good job of becoming civilised if while you proffer western ideas with one hand you take away the fruits of them with the other.'[171] The answer, Huxley suggested, lay in education and empowerment. 'To give the people of Africa a share in the administration of their own territories, and a responsible interest in their economic and social development, is what is wanted.'[172]

Huxley was by no means radical on this issue. He was and remained a believer in liberal Empire, and wanted to see the ceding of only gradual control to black people, accompanied by education and supervision. He wrote in 1940: 'It would be fatal to do nothing about colonies, but it would be equally fatal to try to do too much too quickly.'[173] However, Huxley opposed any restriction of opportunity to black people and attacked white settlers who denied such opportunities. One white settler wrote to Huxley in 1935, challenging Huxley's anti-racism and his apparent coldness to white African colonists as expressed in *Africa View*.

> ...there are points on which a number of people of many years experience will disagree with you, not only settlers, for whom, by the way, you have not many kind words. I was disappointed to reach page 400 before I found any sympathetic reference to them at all and then only a footnote to explain your apparent hostility.[174]

However, the progressive stance on racial discrimination towards black people was more complicated than it appears. As has already been shown in this chapter, Huxley and Haldane were not believers in white and black equality and they continued to articulate their stance on black mental inferiority throughout this period. Their discontent at abuse of black people in Africa was very much focused on white bigotry and the casual manipulation of the race concept as part of a strategy of exploitation. It was not rooted in radical beliefs about racial equality.

As we have seen, Huxley and Haldane were taken to sporadic assertions of white superiority and retained a core belief in racial difference, which held that geographical separation was natural and desirable. For example, Haldane argued that black and white people could not prosper in each other's climates, a view which had been expressed time and again by British racial scientists in the interwar period.[175] Haldane wrote in 1940 that just as Englishmen could not colonise West Africa because of yellow fever, so 'Negroes in England very often die of consumption'.[176] To Haldane, the logical implication of these differing susceptibilities to disease was that black and white should live separately, in their naturally allotted areas:

> It seems to me that where you have evidence of adaptation to environment... it would be desirable to discourage not merely racial inter-breeding, but emigration between the two countries except in so far as it may be necessary for a certain amount of cultural contact.[177]

This was a view which, as we have seen, mirrored government and popular concerns about the suitability of black labour in Britain.

In the context of these opinions it is hardly surprising that Haldane did not become an advocate of increased black presence in wartime Britain. There was here a fundamental difference in his views on Jews and black people which can only be fully explained in racial terms. Haldane was firmly opposed to the persecution of either group in any circumstance. But his advocacy on behalf of refugee Jews would not stretch to supporting black people in Britain for the simple reason that he thought that Jews were racially suited to life in Britain and black people were not. He thus argued:

> A good case can be made out for discouraging immigration of Negroes into Europe, or of Europeans into tropical Africa, since in each case the immigrants are ill adapted...No such case can be made as between the different genetic types (I hesitate to use the word 'races') who have lived in Europe for many centuries.[178]

There is little evidence that Huxley did not share Haldane's lack of enthusiasm for black people in Britain. In fact, his writing in *The*

Inequality of Man indicates that his views closely matched those of his old friend. However, unlike Haldane or Hogben, Huxley did engage with the wartime government on the subject of white–black relations in Britain and the Empire. His willingness to do so was not rooted in any radical thinking on blackness, and it would be a mistake to cast him as a racial radical. In fact, as we have noted, his views on black and white racial difference lagged far behind more militant scientists such as Hogben. Instead, Huxley's engagement with government colonial matters can be better explained by his support for the British Empire. Whilst both Hogben and Haldane were avowed anti-colonialists, Huxley saw a value in trying to inject progressive racial views into the Empire in order to ensure its survival.

Huxley had remarkable access to Britain's policy makers in this period and sporadically used it to instruct the government on racial and colonial matters. In October 1940 he wrote to the Foreign Secretary Lord Halifax with advice about how the colonies should be utilised in the war. In keeping with Huxley's written work in this period, he advised Halifax to use Britain's opposition to German racism as a propaganda weapon to win over hearts and minds in the Empire. To achieve this, Huxley pointed out to Halifax that he should stop using race as a descriptive term even with reference to Germany, in order to gain the benefits of being perceived as anti-racist. He told the Foreign Secretary that he was in any case in error in his use of the idea, that 'in the first place, there is no German race – a German nation, people, culture, tradition, yes; but racially they are even more mongrelized than we'. Huxley pointed out to Halifax that using this vocabulary in itself served to validate the Nazi racial ethos. He reminded the Minister: 'the Nazis have erected racialism into an essential part of their mythology, it is in a sense playing into their hands to use the term!'[179] Halifax seems to have taken this advice on board, indicating Huxley's influence on the presentation of this aspect of the British war effort. He replied: 'I am interested in your remarks about the German race; you are quite right that this is one of Hitler's foibles.'[180]

As for the colonies, Huxley advised Halifax that Britain could gain a valuable propaganda victory by making her colonial administration more liberal and progressive. To this end, he highlighted the need for Britain to focus on 'the interests of local populations, and the wholehearted development, social as well as economic, of the backward tropical areas'. Huxley called for limited but increasing 'self government' including 'high administrative posts' for indigenous people.[181] This progressive colonial management with opportunities for black people would, to Huxley's mind, stand the conduct of Britain in sharp contrast with the racial bigotry of Germany. As such, Britain, through racial progressiveness, could win supporters across the colonised world.

In another attempt to ensure that the correct impression was given out to the Empire, Huxley advised Denis Routh at the Ministry of Information to stop the use of the word 'native' in government broadcasts. Huxley explained: 'In the Gold Coast especially the educated Negroes are very sensitive.'[182] Once again there is evidence that Huxley's advice was taken forward. Routh replied that the suggestion 'which I entirely agree with – is being taken up higher'.[183] Huxley tried to spread his message to other government officials. He sent his 1941 book *Reconstruction and Peace* (written under the pseudonym Balbus) to Anthony Eden and *Africa View* to Oliver Stanley at the Colonial Office. Stanley wrote to Huxley to tell him that he had 'enjoyed reading it enormously'.[184] Huxley and Stanley corresponded about colonial matters throughout the war and in 1944 Huxley was included in a Commission on Higher Education which was sent to assess educational progress in British-controlled Africa.[185]

That Huxley felt the need to communicate with ministers about colonial management indicates that he did not feel that British attitudes towards the Empire were sufficiently conducive to the nation's war aims. The implicit criticism here was that black British subjects were not being treated in the correct way. In one 1940 article, Huxley argued that the chief characteristic of British colonies was 'economic and social backwardness' and that 'Negroes and Malays and Melanasians are human beings, not chattels to be bartered about'.[186] In the context of the war, Huxley extended this argument to contend that black troops were being underused and undervalued. When he wrote in 1943 that the 'achievements of African regiments in the war, both in fighting and in the technical and auxiliary services, have gone a long way to convince sceptics of the immense possibilities of African development' he must have been mindful that, as we have seen, the government did not generally share this progressive view on African troop potential.[187]

In 1944 Huxley received a letter from a Major J.A. Boycott who was commanding British African troops in India. Boycott's views were similar to Huxley's, in that he believed that the British treatment of Africans was 'going to make a good deal of difference to our future relations in the colony'. He asked Huxley to 'put the news about' that his West Africans were 'a damned good force' and that they had 'done well and will probably do better'.[188] Huxley immediately forwarded the letter to Arthur Creech Jones, the Secretary of State for Colonies, keen as he was to have this sort of view taken seriously at the heart of what he knew to be a sceptical government.

Huxley's efforts on behalf of black troops and his progressive writing on the status and rights of Africa's black populations in general, indicates in a small way the same kind of scientific leadership that was evident over the issue of refugee internment. However, it is equally clear that scientists did not give anywhere near the same degree of attention to the issue of anti-black British racism that they did to righting the injustices experienced by the mostly Jewish refugees from Nazism. It is of course true that the great bulk of the energy expended by the scientists on the refugee issue concerned the protection and defence of their own, fellow European academics, who had fallen on such hard times. It is equally true that saving refugees from Nazism brought forth commitment consonant with the level of urgency, largely incomparable to any desire to intervene in state affairs over wartime colonial policy. But these differences do not fully explain the disparity in scientific action.

Ultimately, scientists were more vocal in their leadership as regards the protection of refugees from Nazism because they fully believed in the racial invalidity of Hitler's anti-Semitism and did not fully believe in black potential or the desirability of further black presence in Britain. With the exception of Hogben the radical, they tended to share the kinds of views on black people that resonated in much of British society even if they hated exploitation and sometimes considered, as Huxley did, that advancing the cause of black Britons was a needed political strategy of war. One cannot help but conclude that, prominent public figures as they were, Britain's leading biologists could have done more to intervene on behalf of black people in Britain's war effort, as they did so forcefully on behalf of the refugees. At this time, more than ever, the racial progressives had the ear of the government. Where they saw the need they were more than capable of using their ever-growing influence. On the issue of black war roles they generally did not do so, as for the most part their views mirrored what the government was doing in any case.

More broadly it seems reasonable to conclude that whilst these kinds of issues may, to some extent, have pushed progressive biologists into attempts to lead racial policy, in general they remained fairly well in tune with government and wider society. The underpinning heart of progressive racial theory, its opposition to oppression, its caution and moderation (all the while retaining a belief in racial difference) matched fairly well the sentiments of the British government at war and much of the public and media. Leading progressives were now major public figures and their views on race (still epitomised by *We Europeans* more than any other text) seemed to represent something of the core of British values. But what of the racial conservatives?

The end of the war brought with it a heightened social awareness of the racial atrocities committed by the Nazis. The growing revelation of the Holocaust provided strong ammunition to the progressives' calls for the removal of race from public discourse and brought their cautious and moderate analysis of the concept ever closer to the heart of mainstream British consciousness. How, in this atmosphere, did racial conservatives, still firmly attached as they were to ideas of racial difference and separation, manage to function as scholars and retain any kind of credibility? It is to this question that this chapter now turns.

The perilous mission of the believers: protecting the idea of race in the face of Nazism and the Holocaust

The end of the war brought no urgent changes in outlook or policy from the progressives. As this chapter has already argued, they had mostly taken their end position on race by the early 1940s and the defeat of Nazism did not necessitate any new or dramatic desire for change. For Britain's scientific conservatives the picture was very different, and the end of the war brought substantial new challenges and problems. Not that Britain's scientific racial conservatives significantly changed their views as a result of the conflict. All core aspects of their racial beliefs remained intact, sometimes reworked according to prevailing anti-Nazi winds, but generally unshaken. This continuity is evident in the immediate post-war publications of Arthur Keith and Reginald Gates, both of whom maintained their cases that nations were actually growing races, that 'miscegenation' could undermine the racial quality of a nation, and that races were divided not only by physical differences but by mental ones too.

Keith's 1946 book *Essays on Human Evolution* betrayed anxieties that were commonplace amongst racial conservatives after the war. This text was written during the conflict, and Keith was keen to assert his hostility towards Nazism and uncharacteristically even took Hitler to task for his anti-Semitism. He argued that anti-Semitism 'may be used as a measure of civilisation; its prevalence is a measure of barbarism' and that 'ethically the Hitlerian treatment of the Jews stands condemned out of hand'.[189] Keith's criticism was, though, conditioned by a level of apology for Germany, rooted in his belief that Nazism was a simple manifestation of natural evolutionary behaviour. As we have seen in the previous chapter, Keith believed that modern nations were emerging competing races. In this context he saw Hitler as an 'evolutionist' and 'eugenicist', a man whose outlook was 'somewhat similar' to that of Francis Galton.[190] Hitler's anti-Semitism was presented by Keith as a natural instinct, not as the insane actions of a genocidal maniac, but as the 'uncompromising' stance of an 'evolutionist' whose barbarity was merely 'a reversion to evolutionary behaviour'.[191] War had been necessary, in Keith's mind, to settle what he seems to have perceived as natural racial rivalry:

> Whether we like the present condition of the world, or loathe it, we cannot get away from the fact that war still is one of the main factors at work in shaping the destiny or evolution of human nationalities or races.[192]

It is significant that Keith did not recoil or recant in the face of this evolutionary blood fest. Instead, he continued to assert all of his key racial values. He wrote that the 'only live races in Europe today are its nations' and that their competition and separation were entirely necessary to ensure human advancement: 'if mankind is to be vigorous in mind and progressive in spirit, its division into nations and races must be maintained'.[193]

Gates's first post-war publication made no references to the conflict, and instead focused on the sober scholarly matter of *Human Genetics*. This substantial two-volume tome was focused towards an academic audience and lacked Keith's engagement with contemporary affairs. This may have been a simple issue of scholarly inclination but it could also have been a decision based on a realisation by Gates that his anti-Semitic reading of world politics was best left unpublished. Nonetheless, the nature of the argument in *Human Genetics* set the tone of Gates's post-war scientific career, offering an unwavering defence of the idea of racial mental difference and the principle of racial separation. He told his readers that there existed 'ample evidence' of a relationship 'between body type and temperament' and that 'mental development' was 'clearly not even among the racial groups'. He attacked the ascendancy of scientific theories which emphasised the benefits of racial mixing, noting that the idea of 'hybrid vigour' in man had 'been vastly exaggerated'.[194]

Although the progressive scientists were holding out a tentative hand to new social scientific race studies in this period, Gates was entirely dismissive of the potential contribution that social science could make in what he perceived as a geneticist's field. He was especially hostile towards emerging sociological defences of racial mixing, repeatedly writing in his notes that Ashley Montagu's pro-integration argument in *The Creative Power of Ethnic Mixture* was 'rubbish'.[195] To both Gates and Keith, new arguments which attempted to downplay the importance of racial separation were rooted in unscientific political agendas and conspiracy. Commenting on Otto Klineberg's *Characteristics of the American Negro*, Gates noted: 'The whole book, and many other recent books like it, bear heavily on the idea that all human races must be fundamentally alike, no matter what the evidence to the contrary.' This was, to Gates, 'all part' of what he called 'the Myrdal scheme' to elevate black Americans beyond their racial capabilities.[196]

Gates's and Keith's disaffection with the new treatment of race in social science reflected a more general frustration amongst conservative biologists, who were concerned that the horror of Nazism was leading to a politically driven dismantling of what they still perceived to be an important scientific concept. Hitler's extremism had damaged the respectability of racial science, creating a wave of anti-racist outrage which must have looked like it had the potential to destroy race once and for all. In response, those scientists who wished to retain race as a tangible and worthwhile biological category beat out two paths of resistance in this period. Some, like Gates, chose to dig in their ideological heels and war against a world which they saw as rife with anti-racial conspiracy. This kind of stance, as we shall see below, led Gates to attain something of a pariah status. He, along with like-minded colleagues, was attacked as a Nazi or Nazi apologist across many Anglo-American academic communities. Another strategy, adopted by the liberal-minded leadership of Britain's Eugenics Society, was to rebrand race in the wake of the war by arguing that it remained a valuable concept for study, which had been abused and twisted by Nazism. Whilst the Eugenics Society had more critics than ever in the post-war period, this gentle, almost apologetic approach did achieve some success, not least managing to keep leading progressives like Huxley and Haldane inside the Eugenics movement.

Under C.P. Blacker's leadership the Eugenics Society worked to emphasise the differences between its agenda and the racialism of Nazi Germany. Delivering the Galton lecture in 1945, three months prior to VE Day, Blacker outlined the aims of the British movement set against those of Nazi eugenicists. He contended that there were two essentially different approaches to the idea of eugenics, one 'liberal' and one 'authoritarian'.[197] It was the authoritarian approach, Blacker argued, that had become dominant in Nazi Germany. In contrast, the Eugenics Society was trying to pursue the liberal platform, which was the very 'antithesis' of the German authoritarian position. Blacker thus told his audience of British eugenicists:

> I hope I may be forgiven for elaborating at such length the antithesis between the liberal and the authoritarian outlook on eugenics. I have done so partly because eugenics, theoretically interpreted in terms of racialism and practically applied by authoritarian or fascist methods, has revealed itself as perhaps the most repellent and dangerous manifestation of German National Socialism; and partly because the antithesis has affected eugenic thought in this country and has influenced the policy of the Eugenics Society.[198]

At the heart of Blacker's case lay the idea that the Nazis had corrupted the core values of eugenics, which was intended to help and not harm human populations. Comparing the views of Francis Galton to the racism of Hitler, Blacker noted that it was 'the absence of animus' in the work of the founding father of eugenics which divided him from 'exponents of racism'. After all, Blacker noted, Galton's aim 'was to prevent, not to inflict, suffering'.[199]

In the years that followed the end of the war, Blacker attacked Nazi racial policy time and again, in what seems to have been an attempt to minimise any scope for comparisons to be made between the Nazi agenda and that of the Eugenics Society. However, both the ferocity and frequency of Blacker's forays into this subject area betrayed his awareness that many people in Britain believed that there was more to the comparison than he liked to make out. Blacker concluded an article on Nazi experiments on humans by noting that it was pointless to deny these experiments the label of 'eugenics' in the face of popular opinion:

> Some of you may think that, in view of how irrelevant to what we understand as eugenics are the human experiments I have described, the title of my paper is misleading. You may think that I should have referred to 'so-called' eugenic experiments. Perhaps I should. But the inexorable fact remains that whatever our own views may be, the word eugenics has, through the events I have described, suffered degradation in the eyes of many people and organisations...[200]

In what must surely be seen as an attempt to head off this potentially devastating comparison, the war against Nazism led to changes in the

Eugenics Society above and beyond Blacker's reassuring words. As was noted in the previous chapter, the agenda of the Society had never been primarily focused on racial questions. However, after the war, race became almost a dirty word in the Society, which instead focused its attention ever more closely on contemporary discussions of population.[201] Looking back in 1969, Blacker cited the end of the war as a pivotal moment in this regard. He wrote to J.R. Baker: 'The changes in outlook in the Eugenics Society are, I suppose, a particular manifestation of a general change that has gone on everywhere since 1945.'[202] In this new climate, the Society shrewdly opted to focus on a demographic agenda, which enabled it to retain for the most part a respectable reputation and not fall into the pariah status of racial extremists.

The Eugenics Society was helped in this refocusing by the government's announcement of a Royal Commission on Population in 1943.[203] In Blacker's 1945 Galton lecture he tried to emphasise the closeness of the Society and the government in this process, telling the audience that it had been 'an agreeable surprise...to see how well represented were the Eugenics Society...on the commission and its three expert committees'.[204] There was some truth in Blacker's claims. He himself had been chosen to sit on the 'biological and medical committee' which advised the Commission.[205] Additionally, the Society in its own right was invited to give evidence to the Commission which it did both in written and oral form.[206]

Less than two weeks after the Royal Commission had published its report in 1949, Blacker was once again boasting about the influence of the Eugenics Society in the process. He was delighted that the report had recognised the importance of hereditary factors (described below as 'qualitative considerations') alongside environmental ones in shaping British population development, claiming this recognition as a victory for the Eugenics Society.

> ...the Commission's report is among the few official documents of major importance which has publicly recognised the place of qualitative considerations in social or demographic policy. This standpoint, which is reflected throughout the Commission's recommendations, finds explicit expression in the Report's chapter 15, entitled <u>Differential Fertility</u>; indeed this chapter might appropriately be reproduced exactly as it stands as one of our *Occasional Pamphlets*.[207]

Blacker concluded that 'after a struggle against contrary weather conditions...the winds of public opinion are now blowing in our favour'.[208]

This statement should be read as the propaganda it was, or as a rallying call to the faithful, for Blacker knew full well that much of the public opinion which he described remained hostile to eugenics.[209] He was right about the aspect of the Commission's report to which he gave emphasis. Indeed, Chapter 15 did promote a broadly eugenic agenda. It argued that there was a place in population planning for giving attention to matters of heredity and that the amelioration of economic and social inequalities would not in itself solve Britain's population problems.

> Bad nurture and inadequate education may mask innate intelligence or prevent its full development, but heredity puts a definite limit to what can be achieved in intelligence even with the best of nurture, education and good fortune.[210]

More substantially, some of the overall conclusions of the Commission, notably that Britain's educated and wealthier classes needed to be encouraged to increase the sizes of their families, certainly sat well with the aims of the Eugenics Society.[211]

All in all, however, Blacker's claims concerning the influence of the Eugenics Society on the Royal Commission on Population were based upon his giving exaggerated focus to one part of the Commission's analysis. As a whole the report was, he knew, far more focused on issues of nurture like health and education than on heredity. The Commission's main recommendations focused on an agenda of economic and welfare reforms, designed to help all Britons (not any one selected constituency). Fitting neatly with the radical social aims of Attlee's Labour government, it called for the development of family benefits and the improvement of housing, health and education.[212] This agenda, as Soloway has noted, 'was based upon egalitarian premises that still ran counter to eugenic sensibilities no matter how much [eugenic] reformers acknowledged the importance of nurture'.[213] Even the Biological and Medical Committee, on which Blacker sat, focused much more on issues of nurture than nature in their submission to the Commission, which called for improvements to health care to reduce stillbirths, infant deaths and childlessness.[214]

Furthermore the Commission itself was, by the time it issued its report, only partially in tune with the views of both the government and the public. Having been conceived amid wartime fears of population decline, by the time of its 1949 launch (in the midst of the post-war baby boom) the report seemed less important. The Commission itself concluded: 'the indices which were mainly responsible for the pre-war belief that the

British people was failing to replace itself now point to the opposite conclusion'.[215] Grebenik has highlighted this demographic change as evidence that the Commission's final report did not have a substantial political effect: 'its publication made very little impact on public opinion...was never discussed by the House of Commons and few, if any, of the Commission's recommendations were adopted'.[216] However, Grebenik perhaps overstates this marginality. Despite the baby boom, this was still a period of great concern about the ability of the nation to renew itself and thus maintain its world role. The Commission's fear, expressed in 1945, that they were dealing with no less than 'the ultimate threat of a gradual fading out of the British people' remained relevant in 1949. In this period of real decline in British world power the spectre of undermining demographics still lingered in the thinking of the government as well as that of the Commission.[217]

Blacker's attempt to focus the Eugenics Society onto this social concern was a smart move. Whilst eugenics was not at the centre of the Royal Commission, and the Commission was not at the centre of British policy making, neither the Society nor the Commission were politically irrelevant. The Commission played an important role in shaping political discussions on population, and Blacker worked to see that the Eugenics Society had their say. Through this work, he ensured the continuation of the Society through tough times, even if its message was now more muted and measured on racial matters than some of its members would have liked. Blacker's success (rooted in his own liberal outlook) comes most clearly into view when contrasted with the fortunes of those believers in race who did not adapt to the new post-war climate. The fate of Reginald Gates after the war provides an example of one such scientist.

Reginald Gates left his Chair at King's College in 1942 to take up a visiting research position at the Marine Biological Laboratory at Woods Hole, Massachusetts, where he had studied 30 years earlier for his doctorate. His reasons for leaving the UK are unclear. It seems reasonable to speculate that along with wishing to work in a safer environment where more scientific research could be done, Gates, as a determined anti-Semite, preferred to distance himself from the politically charged anti-Nazi (and pro-refugee) atmosphere of wartime British scientific communities. Gates's departure seems to have protected him from the need to condemn Nazi anti-Semitism like his friend Arthur Keith had done. From the other side of the Atlantic, Gates could and did remain silent on these matters. Whatever his reasons for going to the States, by the end of the war Gates was looking for a new post. In his absence he had been retired by King's in 1945, seemingly without consultation.[218]

Gates initially took a research post at Harvard, before moving in 1946 to take up a visiting fellowship at Howard University in Washington DC, awarded to him so that he could finish research that he was conducting into 'Pedigrees of Negro Families'.[219] Howard, since its establishment in 1867, had been committed to the education of all students without regard to their race.[220] However, in the face of endemic prejudice against mixed-race education it had become in effect a 'Negro University'.[221] By the 1940s, Howard set out to lead the field in 'Negro' studies and thus decided on the appointment of several new experts in 'Negro' affairs.[222] In 1943, the university appointed African history expert William Hansberry. In 1944, Howard financed new black history and social affairs projects led by W.E.B. DuBois and Munro N. Work.[223] It seems likely that the university perceived Gates's interest in 'Negro' families as a similar project. Like Hansberry, Gates was a close associate of Harvard anthropologist E.A. Hooton. He was also warmly supported in his work by Howard's head of science U.H. Ellinger.

Soon, however, Gates's work at the university began to attract criticism. In February 1947, 18 of his academic colleagues signed a petition calling for his dismissal.[224] This petition, addressed to the Dean of Liberal Arts, called for Gates's removal on the grounds that he was teaching racist theory in the university, doctrines 'long since repudiated by objective scientists but associated with Houston Stewart Chamberlain, Gobineau and even Hitler'.[225] The letter writers charged that Gates was guilty of 'violating the principles and ideals of the academic profession', and that he had deliberately ignored 'a substantial body of scientific knowledge in direct contravention of the position he has assumed'. In particular, Gates's opposition to 'miscegenation' between black and white populations infuriated other academics at Howard.[226] In a traditionally black university, such opposition to integration was interpreted as a racist belief in black inferiority, a view that the petitioners considered to be completely unacceptable:

> Every additional day allowed Professor Gates to sully the minds of our students only seems to multiply the humiliation he has already caused us and to further prostitute the traditions of our university. Most of our students are callow but not sufficiently so to accept without protest one who tells them in the classroom that they are inferior.[227]

Responding to the petition in a letter to the Dean, Gates described the behaviour of his colleagues as 'astonishing'. Dismissing the signatories as non-scientists, he accused them of operating a political agenda against his work.[228] In keeping with Gates's general attitude, he seems to have perceived that international Jewish conspiracy was behind the action against him.

As we have seen in the previous chapter, Gates retained a long-standing belief that Jews were responsible for the scientific swing against race that was taking place in Anglo-American scientific communities. The scientific trend calling for the recognition of black racial equality, Gates's 'Myrdal scheme', was to his mind a Jewish conspiracy. He noted in 1944:

> It is a curious fact, which presumably has some devious psychological explanation, that the bulk of biological writers who take the role of apologists for the Negro and attempt to decry the existence of any racial differences are themselves of Jewish stock.[229]

In a letter to the French hygienist and race writer Rene Martial in 1947 Gates was even more explicit: 'In this country there is wide spread propaganda, mostly in the hands of Jews, in favour of unlimited race crossing.'[230]

In the context of these views it is unsurprising that Gates saw the attacks against him at Howard as a manifestation of Jewish racial conspiracy. In response to the petition he wrote to the President of the university, arguing that the action had been 'instigated by Jewish members of your University, who have thus tried to stir up enmity in order to serve their own ends'. He now understood, he wrote, 'the form and methods that Jewish propaganda can take'.[231] After further acrimonious discussion and correspondence, Gates left the university slightly earlier than planned at the end of the term to resume his fellowship at Harvard.

Gates's fall from grace at Howard tells us much about the changes that were taking place in scientific communities after the Second World War. Gates was of course wrong to blame his difficulties on Jewish conspiracy. In reality, the criticism meted out to him reflected two key changes in the study of race after the Second World War. The first of these changes concerned the passing of authority on racial difference from biologists to social scientists, a trend already highlighted in this chapter. The second concerned the increased politicisation of racial study, which made it nearly impossible to proclaim anti-'miscegenation' views or a belief in black inferiority without attracting major professional and social criticism.

As we shall see in the next chapter, Gates was far from the only biologist who was unhappy about social science's newly acquired authority on racial matters. However, his disaffection at Howard illustrates the angst of biologists in this period at its most dramatic. Gates had some basis for his complaint to the Dean, that the staff who had signed the petition against him were not scientists. Of the academics that called for his dismissal there were four historians, five linguists, five social scientists, an anthropologist, a mathematician, a philosopher and an economist.[232] Gates's further assertion, that the evidence cited by the petitioners in order to refute his racial views was not biological but instead research produced by 'a series of anthropological writers', was similarly valid.[233] When the petitioners criticised Gates for 'deliberately ignoring' scholarship, the experts that they listed in support of the allegation were all sociologists, psychologists or anthropologists; there was not a single biologist among them. The petition argued: 'Franz Boas, Ruth Benedict, Margaret Mead, Ashley Montagu, Otto Klineberg and others will serve as adequate documentation of our position.'[234]

Ultimately, Gates was not under attack from within his own discipline but had fallen victim to new social scientific authority. In fact, the Head of Science at Howard, U.H. Ellinger, shared Gates's views and wrote to the Dean in his defence.[235] This is not to say that all biologists adhered to Gates's racist perspective. We have already seen that they certainly did not. However, biologists often did share Gates's belief that he had the right to conduct his research according to his understanding of genetics and physical anthropology. But in the new climate of social scientific research on race, the very parameters of Gates's research agenda isolated him and led to criticism even before his racist conclusions compounded his troubles.

However, Gates's problems at Howard cannot fully be explained in terms of an academic shifting of authority on race. The other key factor which led to Gates's increasing isolation was the heightened politicisation of racial matters in the wake of the war. As we have seen, other scientists (like Blacker at the Eugenics Society) had responded to the fact that the fight against Nazism had rendered the science of race politically contentious on a new scale. Perhaps due his wartime insulation in Massachusetts, Gates had made no such adjustments.

The naivety of his approach at Howard is striking, especially in retrospect. One might have thought that a 'Negro' university was not the best place of work for a white supremacist like Gates. As has already been noted, Gates seemingly agreed to take a position at the university in order to locate some living data for his *Pedigrees of Negro Families*. It was only after the conflict at Howard that he seems to have realised that his decision to use black students as anthropological data could be contentious. A 1950 letter from anatomist Howard Cummins commiserated with Gates on his problems at Howard, concluding: 'It is no wonder that you could not indicate the institutional source of material published in *Pedigrees of Negro Families*, though unlike one of the reviewers I do not consider the omission a fault.'[236] Until the petition was raised, Gates did not think that his attitude towards black people would be a problem at Howard. He was, in his own mind, merely an objective and reasonable scientist, who wanted to develop his biological research into racial difference for the good of humanity. He did not see why it was inappropriate to say that black people were different from whites. This was, for Gates, a matter of scientific logic, which it would be a dangerous folly to deny. He explained his position in these terms to the Dean:

> Some of us, who have been studying racial questions sympathetically for many years, feel that the ameliorating of the social relations between races, which we all desire, can best be attained, not by pretending that racial differences do not exist but by frankly recognising their nature in so far as we can understand them.[237]

Concluding his defence letter to the Dean, Gates re-emphasised his lack of animus towards black students, in a manner which reflected the tone of the Southern segregationists with whom he would soon make new alliances. He told the Dean that his science was a matter of fact not racism. After all, he retained 'the greatest sympathy for the Negro people'.[238]

Gates seems to have been genuinely shocked by his treatment at Howard. Only gradually over the following years did he begin to realise his level of marginality and the extent to which the views that he cherished were now socially and politically taboo. Had he concluded his career ten years earlier he would have probably done so without inviting any kind of major controversy. But in the climate after the war, in the face of US segregation debates and the ascendance of the apartheid regime, Gates lost his reputation as a biologist and gained a new one as a torchbearer for white supremacy.

In the final reckoning it is arguable that the distance between Gates's position on race and that of those scientists described here as progressives was not great and that it was really political (not scientific) divisions which separated the racial theories of Gates (and Keith) from the likes of Huxley and Haldane. Gates was not prepared to use different or softer language to discuss race, or to emphasise the limits of its importance. Unlike his progressive peers, he was not an opponent of Nazism, segregation or apartheid, and saw no good reason to trim his position to the prevailing anti-racial winds. This failure to adapt, as we have seen, brought great social and professional difficulties. By the end of the war, and for the following generation, the progressive position that was born in Britain in *We Europeans* was firmly ascendant, and overt deviance such as that displayed by Gates risked public and professional allegations of racism.

Gates's repeated complaint, that politics was dictating the new anti-racist platform in science, was understandable. Little, it could be argued, had changed in the core beliefs which had driven thinking on race in the natural sciences since the 1930s. Britain's biologists continued to believe that there existed both physical and psychological disparities between different populations, and that these were rooted in both hereditary and environmental causes. However, it would be wrong to describe scientific thinking on race between the 1930s and the 1940s as entirely static. Although scientists were rarely prepared to write unequivocally on the subject, attitudes towards both racial difference and racial mixing had developed. It remained true that none of the leading progressive biologists (with perhaps the exception of Hogben) were consistent in expressing non-racial views, but statements about both racial difference and the deleterious consequences of mixing were less and less heard as the 1940s moved on. This may have been predominantly a change which reflected more the political than the scientific world but it was a change nonetheless.

In British politics, the desire to take up a politically defensible position on race and yet to adhere to older racial beliefs seems to have created tension and discord similar to that evident in science. In the final part of this chapter, these tensions will be explored through an analysis of British immigration policy after the war, specifically of the European Volunteer Workers (EVW) programmes.

Sharing the biologists' uncertainty? The European Volunteer Workers policy after the Second World War

So far this chapter has argued that government policy on racial matters during the war mostly mirrored the ideas of progressive racial scholars and that after the war the progressives' platform was ascendant in natural science. All but extremists like Gates and Keith subscribed to new rules of engagement in racial study, including, as we have seen,

Blacker's eugenicists. This platform, it has been argued, did not signal the end of race. Instead it allowed for a subtle continuation of racial understandings of difference, mitigated by a heightened emphasis on environmental influence and accompanied by a strong hostility towards racial oppression. In the wake of the war, government policy can be seen to have continued to mirror these principles of biological thinking on race to a considerable extent, leading to immigration policy which still saw race as a real factor but was determined to be perceived as anti-racist. The government seems to have operated, like most biologists, with a continuing belief in the racial difference of populations, and a quiet ongoing uncertainty about the potential effects of racial mixing on the nation, along with a strong awareness of the need to deny that there was any racist rationale behind political decisions. Similar to the scientists, political beliefs in race became a guilty secret in this period, lurking in the shadows of ostensibly non-racial policy.

The decision to bring EVWs to Britain after the war was rooted in the already cited fear held within government circles about Britain's declining population, and in more immediate concerns about a shortage of labour in certain industries.[239] To solve these problems, the government promoted labour recruitment schemes across Europe, targeting certain populations which had been 'displaced' during the war. Specific programmes recruited mainly Baltic, Polish, Ukrainian and Italian immigrants for placements in British industry.[240] Before they were accepted, the immigrants were assessed for physical and mental suitability by teams from the Ministry of Labour. Once selected, they were allowed to come and work in Britain though they were, in principle if not always in practice, restricted to certain kinds of employment.

When the Royal Commission on Population finally published its report in 1949 it all but discounted the idea that immigration could be used as a method to maintain the population, describing the prospect as 'impractical' and 'undesirable'.[241] By this time, government schemes had already brought 300,000 EVWs into the UK. The Commission's reluctance to encourage further immigration was not based in opposition to the EVWs, more in a concern that this well of immigrants was running dry. Describing the potential of securing more EVWs for Britain, the Commission noted that there was only 'a temporary source...[which would]... soon be exhausted'.[242] It was true that the reservoir of displaced persons which had fed the EVW schemes was running low but there was no shortage of immigrants in general and it would have been easy to lure additional migrant workers to the UK. It was, however, precisely the prospect of the arrival of other immigrants which the Commission perceived as 'undesirable'.

Understanding the Commission's thinking on the suitability of EVWs in contrast to other immigrants exposes wider governmental rationales about immigration policy in the wake of the war. The Commission argued that 'immigration on a large scale...could only be welcomed without reserve if the immigrants were of good human stock....'[243] Britain, it stated, had a limited capacity 'to absorb immigrants of alien race and religion'.[244] These criteria excluded two major pools of potential labour: Jewish Holocaust survivors and non-European (black) immigrants. The Commission's report did not even consider, in its discussion of potential new reservoirs of immigrants, any source outside Europe.[245] As for Jews, it was presumably they to whom the Commission referred as it argued that an additional necessity was to bring in immigrants 'who were not prevented by their religion or race from intermarrying with the host population and becoming merged in it'.[246]

By arguing that there was a limit to the nation's ability to absorb 'alien' races and that it was important, when considering potential immigrants, to seek out good human stock, the Commission showed some degree of intellectual deference to the previous generation of racial researches in Britain. As we have seen, the idea that racial quality needed to be the determining factor in assessing the desirability of any potential immigrant had been expressed in earlier research into immigration, and had remained a fixture in the more recent writing of Britain's biologists. Indeed, ideas concerning the importance of immigrant racial suitability had been remarkably persistent in British social and scientific discussion, a repeated if unstable theme from the debates that surrounded the 1905 Aliens Act, through inter-war comparative studies of hosts and immigrants, to Haldane's (and other progressive biologists') wartime writing on African inability to prosper in Europe.

Given this intellectual context, it is perhaps unsurprising that the running of the EVW programmes stuck closely to these principles. Officially, however, the EVW criteria did not take account of the nationality or ethnicity of the workers considered. Correspondence between British staff in Vienna and the Foreign Office in London highlighted the egalitarian approach that was nominally to be adopted: 'We note from the Draft Zonal Instruction issued in Germany that there will be no distinction on grounds of nationality.'[247] A week before this note, the Allied Commission for Austria stated that the important task of British agencies abroad was to choose 'productive people...no matter what their race or nationality'.[248] However, perceptions of racial difference did influence the selection of EVWs.[249] The above note by the Allied Commission in Austria itself demonstrated the centrality of group prioritisation in language similar to the Royal Commission. Whilst the productivity of potential workers was to form the guiding criteria of suitability, the note concluded that any immigrants that were to be considered needed to be 'compatible to the British public'.[250] Racial origin, it was further argued, could be the deciding factor in establishing whether certain workers were 'compatible'. 'For example', the note recorded, 'the situation in Palestine, and anti-Semitics, clearly prevented the recruitment of Jews'.

Despite the fact that immigration was not to be limited to any particular nationality in principle, in practice, entire races could be deemed ineligible or at least 'de-prioritised' if it was felt that due to some trait or tendency they would not fit in with other workers within the British economy.[251] A letter from Major-General T.J.W. Winterton (the Deputy Commissioner, Allied Commission for Austria) to Lord Pakenham at the Foreign Office clarified that perceived national and racial groups were to be prioritised in this way.

> The present scheme provides that, although displaced persons of any nationality are eligible, we shall in fact concentrate on each nationality in turn in accordance with a given priority, and shall not pass from one nationality to another until we have exhausted the immediate possibilities of recruitment.[252]

A.W.H. Wilkinson at the Foreign Office argued at the birth of the programme in January 1947 that various nationalities should be prioritised or 'we may find that other countries will have skimmed the cream of the displaced persons, especially the Balts who are undoubtedly the elite of the refugee problem'.[253] Similarly, the Cabinet Foreign Labour Committee in a 1947 meeting urged that Britain should 'bid quickly' or else 'the cream would be taken by other countries and we should be left with the least satisfactory section from which to select'.[254]

Jews, ex-enemy nationals and *Volksdeutsche* were cited as the unwanted groups within the original EVW programme.[255] The records of a conference between officials from the Foreign Office, the Ministry of Labour and the Austrian Embassy outline this decision.

> Some discussion took place on the question of nationalities which could be brought within the field of recruitment and it was finally

agreed that all except ex-enemy nationals, Volksdeutsche and Jews could be included.[256]

The racial categorisation of displaced persons as Jewish and *Volksdeutsche* caused debate, challenge and some discord within government departments. A letter from the Chancellor about the so-called *Volksdeutsche* revealed discomfort at the operating racial rationale: 'To brand many of these people as German because they are of German ethnic origin is tantamount to perpetuating the Nazi nationality myths.'[257] However, even opposition to the exclusion of the *Volksdeutsche* from the programme was often rooted in racial thinking. For example, Earl Winterton defended the *Volksdeutsche*, not by arguing that they deserved equality with other displaced persons, but by asserting that this group needed to be included in the EVW programme as they were 'our best workers and those most likely to make good British citizens'.[258]

Jewish exclusion from the EVW programme cannot be understood without reference to the sustained potency of Semitic discourses in post-war Britain. Ideas about Jewish attitudes to work, the economy and about Jews as citizens, created a popular image of Jewry which often prevented refugee Jews from being perceived as a suitable source of foreign labour.[259] Race though was not the only reason for Jewish exclusion. As was recorded in the note from the Allied commission in Vienna, another issue was the perception that the public would not accept more Jewish immigrants, especially in the wake of Jewish terrorism in Palestine.[260] There were certainly reasons for regarding British anti-Semitism as a potent force in 1947. At the beginning of August, widespread anti-Semitic rioting occurred across Britain, ostensibly as a result of the hanging of two British officers by Jewish terrorists in Palestine at the end of July. Kushner, in detailed analysis of these riots, has argued that the Palestinian attack was but one feature of wider anti-Semitic feeling amongst sections of British society at this time.[261]

However, it is arguable that government willingness to exclude Jews from the EVW programmes was less rooted in fears of public anti-Semitism than in racial thinking at governmental level.[262] It is relevant to question whether voices within government departments utilised fears of rising anti-Semitism as an excuse to prevent Jewish immigration, which was perceived by them as undesirable amid a variety of Semitic racial discourses regarding Jews as workers and citizens.

Within these discourses, Jews were often perceived as inherently urban, commercial and correspondingly as 'work-shy' as regards manual, industrial or agricultural labour. In parliamentary debate in 1947, Major Tufton Beamish (MP for Lewes) argued that the Jewish refugees he had met in displaced persons camps had been 'unwilling to do a job of work'.[263] Whilst Beamish was an extreme voice of anti-Jewish hostility, the image of the work-shy Holocaust survivor permeated mainstream government thinking.[264] Ernest Bevin warned against the danger that Jewish immigration might lead to the 'undesirable concentration of them in the towns'.[265] Underpinning this concern was the idea that Jews would flock to urban centres irrespective of where they were placed within any employment programme.

A pamphlet produced by the Anglo-Jewish Association (designed to counter anti-Semitism) described the prevalence of Jewish images of this kind. The publication cited common grievances against Jews in order to challenge them. 'Anti-Semites complain that Jews live in concentrated areas in a few large towns...[and] sometimes allege that Jews shirk occupations involving heavy or manual labour.'[266] Whilst Jewish worker discourses cannot be said to have been wholly responsible in themselves for Jewish exclusion from the EVW programmes, the perception of Jews as work-shy seems to have combined with the previously noted desire to exclude Jews in order to prevent anti-Semitism. These factors together ensured the rejection of Jewish Holocaust survivors as a potential labour source. As Kushner has concluded: 'Jews had no place in such considerations as they were seen as unsuitable workers and harbingers of anti-Semitism.'[267] Ultimately, these refugees were not able to enter Britain as EVWs. The only programme that would allow their immigration was the Distressed Relatives Scheme.[268]

The racial nature of EVW policy becomes clear in the loosening of restrictions regarding *Volksdeutsche* and enemy nationals. The 1988 House of Commons Report on the Entry of Nazi War Criminals and Collaborators into the UK in the wake of the war has highlighted the eradication of restrictions regarding the admission of these immigrants as the scheme progressed.[269] The report has shown that even known ex-SS combatants were not in the end excluded from labour programmes.[270] Citing pertinent government records from the period, the report argued that many within the British government saw EVW immigration purely in terms of providing a labour solution and as such were only interested in recruiting the most racially able workers, irrespective of their past loyalties. Bevin, it is asserted, was looking 'purely to secure the most rapid solution of the labour crisis in the UK'.[271] Likewise, the

Lord Chancellor is cited in the report as making the following remarkable statement about the possible immigration of war criminals:

> I am willing to risk their [EVWs] being Nazis – and I think they probably are – so long as they are highly skilled technicians who will teach our people something which they primarily did not know.²⁷²

These findings raise important questions regarding the place of Jews in the EVW schemes. Having initially rejected the idea of *Volksdeutsche* and ex-enemy national immigration on the grounds of feared public disquiet, the government was not ultimately dissuaded from allowing these groups to come, even where there was evidence of their complicity and participation in Nazi war crimes. This suggests that the government were prepared to challenge public prejudices in order to solve the labour crisis.²⁷³ In this context, it is difficult to explain Jewish exclusion from the EVW scheme in terms of a governmental fear of rising anti-Semitism. As significant as anti-Semitism may have been, it was surely not as prevalent or vociferous as hostility towards ex-Nazis in the wake of the war. It seems more pertinent to argue that fears of anti-Semitism were utilised as an excuse to exclude potential Jewish immigrants by members of the government who subscribed to Semitic discourses concerning Jewish worker suitability. Cesarani has concluded that it was indeed racial criteria which set and sustained the EVW policy on eligibility. In explaining why it was that the government were prepared to 'exert' themselves in promoting some workers even in the face of public hostility, he has concluded:

> In considering why East Europeans were deemed worth this exertion, but Jews, Blacks and Asians were not, it is all but impossible to avoid the conclusion that racism was at work.²⁷⁴

Ultimately, Cesarani has argued that the EVW programmes were 'shot through with racist assumptions about "good human stock" and "assimilability"'.²⁷⁵ Regarding the consequences of this approach, he has asserted: 'In the end, it would benefit Balts, Ukrainians and ethnic Germans; Jews, Blacks and Asians would be the victims.'²⁷⁶ If the EVW programme was, after all, a racially informed immigration policy, it is significant that its racial agenda was operated in such a covert manner. The secrecy behind decision making indicates a certain shame about thinking racially, leading political leaders into a practice of denial and mitigation which resonates closely with what was going on in British science.

It has been noted in this chapter that in nearly every case Britain's biologists, like their political peers, retained a belief in racial difference in this period but that the word 'race' was hardly ever used in their writing because they did not want to give legitimacy to Nazi or any other state racism. Nonetheless, British scientists continued (using the language of 'ethnic groups', 'populations', 'cultures') to uphold the idea that population groups were inherently biologically different, both in terms of their physical and mental make-up. Most scientists would mitigate race in any way possible, even discard the term, but they still ultimately believed in it.

British immigration policy developed in a similarly ambiguous atmosphere. Not only, as we have seen, did EVW policy fit neatly with the racial aspects of the findings of the Royal Commission on Population, but it also mirrored wider currents in the thinking of Britain's biologists. Like scientific work in this period, immigration policy was keen to hide any overt racial intentions. Just as Britain's scientists used every available mitigation to hide the racial core of their reasoning, so immigration policy was craftily constructed to hide its real intentions. On the whole, Britain's scientists, and the society from which they emerged, did not lose long-held racial beliefs in this period. In the wake of Nazism and on the eve of apartheid and the US Civil Rights movement they did learn to be ashamed of them. Race had become politically beyond the pale but it had not been dismissed as a biological concept. Scientists and politicians alike still believed that questions of racial difference and racial mixing were of considerable importance.

4
Race on the Retreat? The 1950s and 1960s

Stepping back from society: the progressives, Lysenko and Soviet science

The war against Nazism graphically highlighted the danger of mixing science and politics. Facing Nazi racial violence, Britain's scientific conservatives had often felt driven to retract, revise and refocus their analysis of the concept of race in recognition that ideas, not too distant from their own, had been corrupted into the state religion of the most terrible of regimes. But in the emerging Cold War atmosphere of postbellum Europe, it was not only the scientific right that was forced onto the defensive. Britain's leftist scientists were soon similarly cowed, faced with the brutality of the Stalin regime, the onset of the Cold War, and specifically with Soviet championing of the agronomist Trofim Lysenko.

Lysenko had made his name in the USSR's media in the late 1920s as a practical peasant scientist who had developed simple yet iconoclastic techniques to increase Soviet grain yields.[1] Lysenko's method involved 'vernalisation', which he defined as the pre-treatment of seeds to specific environmental conditions, prior to planting. In this way, Lysenko argued, it was possible to change the growing potential of any crop. This belief in genotypical transformation by environmental means ran counter to the core values of geneticists both in the USSR and the West. However, it carried political implications which were warmly received by Stalin's government and many Soviet agricultural workers.[2] For one thing, Lysenko was not a resident of any academic ivory tower, but was instead a worker scientist interested only in the practical applications of his research. For another, his giving of primacy to environmental factors sat well with Communist political ideology, where genetic theory and the accompanying notion of heredity were regarded with suspicion as practically irrelevant and bourgeois, even as Fascist and racist.[3] By the mid 1930s Lysenko had become a leading light in Soviet science despite hostility from leading Soviet geneticists.[4] As one scholar has noted regarding Lysenko's rise to prominence: 'Lysenko understood that the regime wanted the same things from biology that it wanted from every other sector of Soviet activity: ideological conformity and increased output.'[5]

Many of Britain's leading progressive biologists in this period were outspoken supporters of the Soviet Union. Despite his self-identification as a humanist and rationalist (and not a socialist) Julian Huxley was one such scientist.[6] After the USSR had joined the war Huxley wrote an article on the importance of Anglo-Soviet cooperation.[7] This article reflected a long-term friendship, which had been disrupted by the Nazi–Soviet Pact and the Soviet invasion of Finland. Previously, Huxley had been the Vice President of the Society for Cultural Relations between the Peoples of the British Commonwealth and the USSR and had agreed to sit on the editorial board of the *Anglo-Soviet Quarterly*.[8] In 1945, Huxley visited the USSR and was, according to *Time Magazine*, 'enormously impressed with the Soviet attitude towards science'.[9]

Haldane and his then wife Charlotte had visited the USSR in 1928 at the invitation of the plant geneticist Nikolai Vavilov and had been amazed by the progress of the country.[10] By the time Haldane became a card-carrying Communist in 1937 he was a firm Soviet supporter as was Joseph Needham.[11] Both men were convinced that Stalin was in the process of organising his nation according to a scientific and rational ideology which would give the USSR the edge over her international rivals. Haldane was still prepared to argue this case in 1961, claming that the Soviet Union had an advantage 'over the USA, Britain, France, and so on...[because of its] more scientific world view'.[12] Lancelot Hogben had a more critical attitude. Like the others, he was impressed with Soviet scientific progress in the interwar period and sympathetic towards Marxist ideology.[13] However, a short stay in the Soviet Union in 1940 had convinced him that the regime was both oppressive and dangerous.[14]

The rise of Lysenko posed a difficult challenge to Britain's progressive scientists. All the men described here as having certain degrees of Soviet sympathies were also avowed geneticists. In this atmosphere, it became increasingly difficult to reconcile their political warmth to the USSR with their scientific beliefs. To Huxley and Hogben, who saw themselves as loosely left wing, it was easier to distance themselves from Lysenko. For Haldane it was much harder. As a dedicated and active

Communist Party member the rise of Lysenko put him, as Werskey has noted, 'under severe emotional and intellectual strain'.[15] At the early stages of the controversy he seems to have tried to sit on the fence, to defend some aspects of Lysenko's approach and criticise others. In one 1940 article Haldane thus wrote: 'I have little doubt that he [Lysenko] has gone too far in some directions, but it is important to see what there is of value in his criticism of orthodox genetics.' After all, Haldane noted, 'scientific pioneers are not infallible'.[16] Later, in 1946, Haldane made a similar defence which outlined his specific objections to Lysenko's thinking along with his case for continuing support:

> Lysenko holds a number of opinions on biological questions, with some of which I agree. I disagree with many of his opinions on genetics, though I think that his criticism of earlier opinions was sometimes correct.[17]

At this stage, Haldane clearly still believed that Lysenko was achieving something with his strange science which merited his position and justified his errors. He concluded: 'Lysenko is approaching the truth by a roundabout method.'

Two events in the summer of 1948 helped to iron out the ambiguities and apologism which had characterised much of the progressive response to Lysenko and Soviet science. The first of these was Stalin's elevation of Lysenko to the forefront of Soviet biology and his corresponding purge of orthodox geneticists in the USSR.[18] Until this point, it had been possible to argue that the Soviet Union was pursuing two intellectual paths in a spirit of scientific openness. After 1948 it was difficult to dispute that debate was being stifled and good science suppressed. Haldane began a slow withdrawal from the British Communist Party which continued to support the Soviet position. Although he remained reluctant to speak out on Lysenko or Soviet injustice, his actions revealed his real opinion.[19] Huxley's obituary of his friend in 1965 records Haldane's view:

> ...his scientific integrity was too much for him when Lysenko with his bogus theories... was praised and honoured by Stalin and the Party in Russia (and consequently by the Communist Party in Britain): and he resigned from the Party.[20]

The impact of Lysenko's ascendancy on Britain's scientists was compounded by another major incident. In August 1948 many of the cream of Britain's left-leaning intellectuals were invited to attend a conference in Breslau, just renamed as the Polish city of Wroclaw.[21] The conference was instigated by M. Boresza, a Polish politician and publisher, on the theme, 'Des Intellectuels pour la Paix'. The idea was to build cultural bridges between the emerging Eastern and Western blocs.[22] In practice, the conference served instead to reveal to Britain's leading scientists some of the realities of the new Cold War atmosphere in science. Britain was represented in Wroclaw by Julian Huxley, Joseph Needham, the mathematician Hyman Levy and physicist John Desmond Bernal. These scholars were joined by other leading intellectuals including Kingsley Martin from *The Spectator* and the historian A.J.P. Taylor. There were also delegations from the USA and France including the painters Leger and Picasso, the poet Éluard and the atomic physicist Frédéric Joliot-Curie.[23]

Instead of finding a conference on international peace, the Western delegations in fact fell into something of a Soviet trap. Once in Wroclaw they were treated to a series of diatribes about the moral bankruptcy of Western society. Huxley recorded that half the speeches made 'were attacks, often violent, of one side on the other'.[24] He refused to sign the manifesto drafted at the conference and left, like the great majority of Western delegates, in disgust.[25] In the conference's wake, Huxley wrote prolifically on the dangers and deficiencies of Soviet science (mostly focused on Lysenko) displaying a new resolve to tell the world what he had found out in Wroclaw.[26] In particular, he and others at the conference had learned that Soviet scholars were now overwhelmed by the demand that they parrot a political line and had lost, because of politics, their scientific experimental integrity. Huxley thus recorded that the meeting had been both 'distressing' and 'valuable'.[27] Albert Einstein wrote to Huxley in the wake of the conference:

> Under the prevailing circumstances it is impossible for the intellectuals of the East and the West to work together for peace and intellectual freedom. I am convinced that our colleagues on the other side of the fence are wholly unable to express their real opinions.[28]

Ultimately, the Wroclaw conference and the Lysenko affair (as striking examples of the new Cold War atmosphere) did for many of Britain's progressive scientists what Nazism had done for the conservatives. It taught them the hard way that there were good reasons to keep science and politics well apart. Again, the issue of race was at the heart of this lesson. Haldane and many other leftists believed that it was Nazi racism that had driven Soviet science into the arms of Lysenko in the first place, as new explanations of life and progress were sought that were environmental, not based on heredity. Asking permission from his employers to go and lecture in Prague in 1948, Haldane explained the importance of his trip in terms of the need to rescue Czech science from a Nazi-inspired flight from genetics:

> There has been a tendency in Czechoslovakia, since at least as early as 1945, to reject certain genetical ideas altogether, especially in relation to man, as a reaction against the Nazi racial theories. Recent controversies in the Soviet Union have reinforced this tendency. It seems to me important to combat it.[29]

Supporting Haldane's understanding of the turn in Eastern bloc science, Werskey has noted the tendency of the Lysenko camp to play off 'the practical and democratic orientation of their socialist science against the sterile and racist nostrums of capitalist pseudo-science'.[30] This behaviour was indeed evident at Wroclaw, where the majority manifesto accused the Western scientists of inheriting 'fascist ideas of racial superiority'.[31] Jones has concluded: 'Lysenkoism was often depicted as an attempt by the USSR to escape the racist implications of research in heredity.'[32]

The progressive scientists, who had been so political for so long, thus became in the Cold War belatedly wary of mixing science and politics. Faced with the new situation 'a new, highly apolitical generation of academic researchers was coming to the fore'.[33] Britain's leading progressives, who had been instrumental in leading public thinking on race before and during the war, now wanted the subject to drop back into the scientific arena. As Jones has noted: 'Lysenkoism acted as a final signpost on the road to political disillusion.'[34] This desire would enable in time some renewal of friendships and alliances with Britain's racial conservatives who had, since Nazism, been desperate to rescue what they perceived as the legitimate study of race from the glare of political attention. However, as we shall see, the determination of some racial conservatives to attach scientific racial study to new political issues (like segregation and apartheid) in the end ensured that old animosities would mostly remain. More broadly, the new passivity of the progressive biologists also served to ensure that anti-racial political leadership passed once and for all to social scientists, who still wanted to operate within the political sphere on the issue. Nowhere was this new state of affairs more evident than in the issues surrounding UNESCO's first statement on race in 1950.

British biologists and the UNESCO statements on race

The drive for UNESCO to produce a statement on race began in earnest in 1948 when the United Nations' Economic and Social Council advised UNESCO of a UN interest in developing educational material which could serve to prevent racism and protect minorities.[35] It was with this request in mind that the fourth general conference of UNESCO approved a resolution calling on its Director General:

1. To collect scientific materials concerning problems of race;
2. To give wide diffusion to the scientific information collected;
3. To prepare an educational campaign based on this information.[36]

The Director General of UNESCO in 1949 was Dr Jaime Torres-Bodet, a Mexican poet who had taken the post over in 1948 from Julian Huxley, who had held it since the organisation's beginnings in 1946.[37] Torres-Bodet commissioned the head of UNESCO's Social Sciences department, the Brazilian anthropologist Dr Arthur Ramos, to carry forward the resolution. Ramos in turn convened a committee which gave its report, the UNESCO 'Statement by Experts on Race Problems', in July 1950.[38] The committee of experts created by Ramos was made up of social anthropologists and sociologists. The UK was represented in the committee by Morris Ginsberg, Professor of Sociology at the London School of Economics. Of the ten people on the committee, there was not a biologist amongst them.[39]

In the context of racial study in the years preceding the Statement (and given that UNESCO's founding Director General had been Huxley) it seems remarkable that a group set up to collect and diffuse 'scientific materials concerning problems of race' did not include any biologists. Perhaps Ramos's own background in anthropology, and that of Robert Angell, the Michigan sociologist who replaced him after his death in 1949, shaped the committee's composition. However, the committee's make-up also reflected a larger international trend, a passing of authority on racial matters from natural to social science.

It is arguable that UNESCO itself and specifically its 'Statement on Race' grew out of an international reaction to Nazi racism. When the UNESCO commission moved to Paris in November 1946 and was lodged in the same hotel which had previously housed Nazi oligarchs, Huxley noted that it 'visually symbolized the transition from war and racialism to peace and cultural cooperation'.[40] Similarly, the convening of the UNESCO race committee in Paris's UNESCO house (which had been the Nazi headquarters during the occupation) led the sociologist Ashley Montagu to record that 'only if our deliberations had taken place at Auschwitz or Dachau could there have been a more fitting environment to impress upon the Committee members the immense significance of their work'.[41] To Montagu, the Statement was a fitting and direct response to the Holocaust, needed to refute the horror which had been perpetrated in the name of science:

> In the decade just passed more than six million human beings lost their lives because it was alleged they belonged to an inferior race. The horrible corollary to this barbarism is that it rested on a scientifically untenable premise. On this the scientists of the world are agreed. And through an agency of the United Nations a group of them have gone on record to clarify the whole concept of race.[42]

It would be an exaggeration to say that this atmosphere made biological experts on race *personae non gratae*. After all, and as we have seen, Britain's most famous biologists led the charge of opposition to Nazi science. Nazism had nonetheless tainted the idea that biology had a role to play in racial studies. In what can perhaps be seen with hindsight as throwing the baby out with the bathwater, biologists were not included in UNESCO's 1949 list of racial experts. Instead, the anthropologists and sociologists chosen by Ramos put together a statement which refuted Nazi racialism in a way which biologists could not and would not have done. With a polemical clarity, the 1950 UNESCO Statement on Race dismissed biological ideas of mental racial difference and in doing so pushed anti-racial study into a new era.[43]

The absence of biologists from the creation of the first UNESCO Statement did not lead to a Statement without biological foundation. In fact, most of the argument of the Statement was rooted in an exposition and articulation of race which was essentially biological. Articles One to Four gave themselves over to defining race in a scientific sense, explaining the true biological meaning of the term, referred to in Article Five as 'the scientific facts'.[44] A race, the Statement argued, could be defined in a biological sense 'as one of the group of populations constituting the species *Homo sapiens*'. These populations, stated Article Four, were 'characterised by some concentrations, relative as to frequency and distribution, of hereditary particles (genes) or physical characters, which appear, fluctuate, and often disappear in the course of time by reason of geographic and/or cultural isolation'.[45]

Not only was race defined in the Statement in biological terms, but biology was also presented as the discipline that offered real facts on the subject, in contrast to social misinterpretations. Article Fourteen thus noted that 'the biological fact of race and the myth of "race" should be distinguished' and cited Darwin in support of the point.[46] Much of the language in the rest of the Statement similarly displayed deference to progressive biological research from the previous generation. In particular, the statement in Article Six which suggested that the term 'race' should be substituted in 'popular parlance' with the idea of 'ethnic groups' probably came directly from *We Europeans*.[47]

However, in one crucial way, the Statement went further than the progressive biologists had ever done; it unequivocally dismissed the idea that there were racial mental differences between populations. The Statement outlined that whilst an anthropologist might try to classify human groups in terms of their physicality, he 'never includes mental characteristics as part of those classifications...tests have shown essential similarity in mental characters among all human groups'.[48] Article Eleven argued that there was 'no definite evidence...[of] ...inborn differences between human groups' as far as temperament was concerned. Article Twelve similarly contended: 'personality and character...may be considered raceless'.

This dismissal of racial mental difference may explain the absence of biologists from what seems to have been a statement about biology. The creators of the Statement were fully aware of the contribution that had been made in previous decades by progressive biologists to the intellectual deconstruction of race. They, however, wanted to go further. In the first UNESCO Statement, race was presented as an unstable physical phenomenon which carried no meaning about the inherent psychology of different populations. Mental differences were, according to the Statement, culturally created:

> The scientific material available to us at present does not justify the conclusion that inherited genetic differences are a major factor in producing the differences between the cultures and cultural achievements of different peoples or groups. It does indicate, however, that the history of the cultural experience which each group has undergone is the major factor in explaining such differences. The one trait which above all others has been at a premium in the evolution of men's mental characters has been

educability, plasticity. This is a trait which all human beings possess. It is indeed, a species character of *Homo sapiens*.⁴⁹

To the creators of the Statement, mental differences between ethnic groups needed to be considered as an aspect of social science. In this way, cultural differences were explained as a matter of life experiences, without any tangible biological basis. The Statement noted: 'given similar degrees of cultural opportunity to realize their potentialities, the average achievement of the members of each ethnic group is about the same'.⁵⁰ Montagu elaborated on this idea in his annotated notes. Commenting on Article 11 (on the inheritance of temperament), he argued: 'When "the expansive and rhythm-loving Negro" is brought up in England he becomes as composed and phlegmatic and as awkward rhythmically as is the average Englishman.'⁵¹

This declaration of intellectual and psychological human similarity was too much for most natural scientists to stomach and led to a backlash against the UNESCO Statement. Whilst Jones is no doubt right to stress the importance of the Cold War atmosphere in shaping objections to this kind of egalitarian science, the protest against the first Statement was also rooted in scholarly and professional discord.⁵² The protests of biologists and physical anthropologists revealed their reluctance to confirm the idea that there was no such thing as mental racial difference. It also reflected their exasperation at the passage of expert status on race out of their hands, into sociology and cultural and social anthropology.

The widespread dissatisfaction of biologists with the 1950 Statement is epitomised in the reaction of Julian Huxley. Some historians have misread Huxley's role and response to the Statement, assuming that he was supportive and that he drove the process forward. For example, Barkan has mistakenly claimed that those 'scientists who were involved in the pre-war years with the anti-racist campaign, such as Julian Huxley, J.B.S. Haldane, L.C. Dunn, and Otto Klineberg, were leaders in the new enterprise'.⁵³ Kevles too has placed Huxley in the pack of UNESCO Statement leaders, claiming that the Statement was 'the product of an internationally distinguished effort – the drafters and commentators included Otto Klineberg, Hermann Muller, and Julian Huxley'.⁵⁴

This historical slip is probably due to UNESCO's inclusion of Huxley's name at the bottom of the Statement, where he is listed among many other scientists as having submitted criticisms upon which the text was then 'revised'. Huxley may or may not have submitted corrections in this way, but there is strong evidence that he did not support the publication of the UNESCO Statement, which he believed to be inaccurate and scientifically invalid.

Huxley first saw the Statement in January 1950, when it was sent to him by Robert Angell. Huxley immediately found fault in the document. He did not object to the idea of attacking racism and the false science that was used to justify it. He had traditionally been, as Kevles and Barkan have both noted, at the forefront of such attacks over the preceding period. However, the confident assertions of mental racial equality which dominated the Statement were too strong for Huxley. Commenting on the draft, he responded to Angell: 'as it stands, I think it most unsatisfactory and indeed unscientific in making sweeping but unproved assertions as to racial differences being entirely due to environmental and social factors'.[55] The key word in Huxley's objection was 'entirely'. In his anti-racist writings he had always mitigated racial mental differences by stressing the interaction of inherent and environmental factors in shaping human development. It would have been correct, he told Angell, to stress 'the great role' which environment played in shaping human difference but not to suggest (as the Statement did) that matters of development and difference were 'entirely' environmental.

To Huxley, the Statement's unequivocal (non-racial) explanation of difference was a betrayal of biological truth. Instead, he cautioned, UNESCO should have recorded in the Statement that there was in fact a scientific 'presumption that the major races of man will have genetic differences in mental and temperamental characters as well as in physical characters'. Huxley proposed that instead of hiding this reality, UNESCO should 'stress the need for further research into this important but difficult subject'. To let the document go out in the form which he had read, he told Angell, 'would be bad for UNESCO's reputation, and bad for the causes that UNESCO has at heart'.[56]

At the core of Huxley's objection was a belief that the correct men had not been selected by Ramos for the committee. The man chosen to represent Britain was to Huxley's mind a non-specialist. More broadly, Huxley was put out that biologists had not been used to draft a statement which addressed the 'genetical basis of race'.

> I note the eight experts responsible for the statement. I do not recognize the name of any geneticist, or indeed biologist among them. This seems to me very extraordinary. Morris Ginsberg I know and appreciate very much, but he has no special knowledge of genetics

which would entitle him to be a party to such sweeping statements as to the genetical basis of race.[57]

Ultimately, Huxley's concerns were not addressed. UNESCO released the Statement to a volley of scientific criticism.[58] Alfred Metraux, from the UNESCO Department of Social Sciences, confided to Huxley that he was 'particularly distressed that… [Huxley's]…suggestions and warnings went unheeded'. The Statement had been met, he noted, 'with a strong current of adverse criticism from physical anthropologists and biologists, especially in the United Kingdom'.[59]

This criticism from biologists mostly did not concern the Statement's anti-racist motives. Like Huxley, the great majority of Britain's physical anthropologists and geneticists were keen to step/remain on the anti-racist bandwagon, only they did not want to abandon what they perceived as scientific accuracy in doing so. Commenting on the response of biologists to the Statement, Banton has thus observed: 'while not rejecting its general spirit or its main conclusions, [they] believed that it went beyond the scientific facts'.[60]

One of the clearest examples of the British scientific reaction to the UNESCO Statement can be seen in the pages of *Man*, a monthly journal of anthropology published under the direction of the Royal Anthropological Institute. The editors of *Man* responded to UNESCO's Statement by writing to *The Times* on 15 August 1950. In this letter, they broadly welcomed in principle UNESCO's effort, expressing 'cordial agreement with the purpose and essential thesis of the document'. However, the letter went on to note that the radical renunciation of race in the Statement was something 'to which very few anthropologists anywhere would yet venture to commit themselves'.[61]

This was a typical response from mostly physical anthropologists who, like biologists, were not prepared to see race analysed as an entirely social phenomenon. In the months that followed, *Man* included the letters of numerous scientists who wished to comment on UNESCO's Statement. Matching the editorial position, most contributors welcomed the tone and agenda of the Statement but stopped short of validating the science behind it. H.J. Fleure, for example, focused on the Statement's value in disproving false racial doctrines. He noted: 'One is glad that the UNESCO document takes a firm stand against this nonsense.'[62] In private correspondence, however, Fleure was critical of the extent of the Statement's anti-racial position. He confided to C.P. Blacker: 'I liked parts of the UNESCO statement but some of it was just hot air from Ashley Montagu – An American of East European origin full of complexes.'[63] Similarly, the anthropologist Kenneth Little wrote that he supported 'most heartily the essential thesis of the document' but felt that it had gone beyond scientific evidence in its determination to make an admittedly worthy ideological point.

> Certain of its statements and conclusions suggest a philosophical or ideological doctrine rather than a 'modern scientific' one. For example, what evidence is there for any scientific belief that man is born with biological drivers towards universal brotherhood and cooperation and how is it adduced?[64]

The contributors to *Man* were far from united in their specific views and criticisms of the Statement.[65] There was, however, widespread agreement that UNESCO had not achieved a consensus, had gone too far as regards its claims of mental racial equality, and was too close in its analysis to Soviet egalitarian science. As Huxley put it in another letter to UNESCO, the social anthropologists who had drawn up the Statement were not led by science but 'coloured by wishful thinking'.[66] In an attempt to fix the problem, the editors of *Man* liaised with UNESCO's Alfred Metraux over the possibility of creating a revision, a new Statement this time written in consultation with geneticists and physical anthropologists. Under the pressure of criticism, Torres-Bodet commissioned a redrafting from 'a new panel of experts to discuss exclusively the biological aspects of race'.[67] To help achieve this, Metraux asked Huxley if he could suggest some physical anthropologists, geneticists and a taxonomist 'of indisputable reputation, who would, by the consensus of opinion, be truly representative of English biological sciences'.[68]

Huxley wrote a list of scientists who he felt would be suited to the task.[69] Of the several names that he put forward only one, the Birmingham anatomist Solly Zuckerman, was actually co-opted onto the panel. The others on Huxley's list were rejected by Metraux for being too conservative on racial matters. Huxley himself conceded that, of his nominees, C.D. Darlington could 'be rather extreme' on race and R.A. Fisher was 'rather inclined to take an extreme eugenic point of view'.[70] But Huxley probably recommended these conservative figures deliberately to ensure that the second UNESCO attempt at a statement on race did not make the same mistakes as the first. He explained his conservative selections to Metraux in this way:

> I do feel strongly that you should have someone on it capable of putting the point of view of modern evolutionary genetics...If you

do not have a good theoretical geneticist you are likely, I fear, to get further sweeping statements about the non-existence of any racial mental difference.[71]

The committee that emerged from these deliberations was a compromise between the kind of conservative panel called for by Huxley and the radical aspirations that had driven the first Statement, epitomised in the continued presence on the second panel of Ashley Montagu. The UK was represented on the new committee by Zuckerman and J.B.S. Haldane along with the anthropologists J.C. Trevor and A.E. Mourant. Whilst the choosing of these men definitely represented selective picking from the liberal end of Britain's scientific elite, the presence of such prominent figures, alongside similarly impressive international colleagues, ensured that both genetics and physical anthropology were well represented beyond dispute in the new committee. The editors of *Man* noted that the new panel 'was excellently balanced in point both of specialisms and of nationalities'.[72]

The second UNESCO Statement was released in July 1952. Despite Montagu's later claim that there was 'substantial agreement' between this Statement and the first, the new panel in fact produced a document which was noticeably different.[73] Without doubt, the first Statement's aims, to discredit the racist policy which had driven Nazism and attack the idea of racial superiority and purity, remained intact in the second. In L.C. Dunn's record of the meeting of the second panel he wrote that there had been:

> no delay or hesitation or lack of unanimity in reaching the primary conclusion that there were no scientific grounds whatever for the racialist position regarding purity of race and the hierarchy of inferior and superior races to which this leads.[74]

Confirming Dunn's analysis, the claims of the first Statement, that nations were not races, that racial oppression was wrong in every instance, and that there were no grounds for believing in racial hierarchy, were repeated almost word for word in the second.

There were though important if subtle differences between the two Statements. The first Statement's claim that race was in practical terms a 'social myth' more than it was a 'biological phenomenon' found no place in the second. The far-reaching (and difficult to demonstrate) assertions made in the first Statement, claiming that man was 'born with drives towards cooperation' and that individuals and nations 'fall ill' if this drive is not nurtured, were also removed. Most important to this analysis, the idea that there might be scientific validity to the idea of mental differences between races crept back into the second Statement, albeit couched in careful and moderate language.

Where the first Statement had claimed that anthropologists 'never' included mental characteristics in their classifications of man, the second Statement only asserted that 'most anthropologists do not include mental characteristics in their classification of human races'.[75] This was an important if seemingly minor difference as, through the use of 'most' instead of 'never', the second Statement left the door open for the study of mental racial difference where the first had firmly shut it.

The reason for this difference seems to have been that the framers of the second Statement believed that there was some value to investigations of differential racial mentality, beliefs which stood directly in contrast to the first Statement's bold assertions of 'essential similarity'. However, the language of the second Statement did not aim to directly undermine its predecessor. It did not demand recognition that races were mentally different, but instead recorded that variations between groups were 'very slight' and were not 'a major factor' in producing the cultural differences which existed between different peoples.[76] Despite this moderate language, the apologetic and understated tone cloaked a demonstrable difference in scientific opinion in the second Statement.

In a substantial if subtle departure from its predecessor, the second Statement argued that it was possible (though it conceded unproven) 'that some types of innate capacity for intellectual and emotional responses are commoner in one human group than another'. Having made this assertion of potential 'innate' mental difference, the writers immediately conditioned it by claiming that it was certain 'that, within a single group, innate capacities vary as much as, if not more than, they do between different groups'.[77] If this caveat was designed to undermine the idea that one race could be vastly superior to another, the sentence that preceded it seems to have been included in order to discredit the opposite position; the idea that all races were, in terms of innate qualities, equally endowed.

The Statement noted that whilst scientists knew that 'mental diseases and defects' were transmissible through generations, they were 'less familiar with the part played by heredity in the mental life of normal individuals', again seemingly leaving the door open for future studies of racial mental difference. They were, however, certain that everyone was 'essentially educable' and that 'intellectual and moral life' was 'largely conditioned' by 'training' and 'social environment'. This conclusion, whilst seeming to endorse the claims of the first Statement, clearly did not go as far, leaving the issue of mental racial difference as something of a grey area. In doing so, it better represented the perspectives of Britain's leading progressive biologists who were in this period as keen not to prematurely disregard the study of race as they were to ensure that it was not manipulated into misanthropic political projects.

Metraux had always planned to send the second Statement out from the committee room to gain the widest possible scientific criticism prior to publication. However, in the wake of the drafting process, conflicts soon emerged from within the UNESCO committee about when the Statement should be considered ready for dissemination. When Ashley Montagu released the provisional script to the *Saturday Review of Literature* (where it was published on the first of September 1951) it led to a volley of criticism from the British representatives on the panel. In a joint letter to *Man* in May 1952, Haldane, Mourant, Trevor and Zuckerman cried foul about the publication and made it clear that they did not support it.

> We are not prepared to subscribe to the document printed, without the agreement or knowledge of the Committee, in the Saturday Review of Literature of I September, 1951, and, whatever the motives which prompted its publication may have been, we can only deplore this as detrimental to the efficient conduct of UNESCO's campaign against unscientific notions of race.[78]

Montagu responded that he had acted with UNESCO's blessing, claiming that the organisation's 'mass communication unit' had in fact typed the Statement and that Metraux himself had sanctioned 'immediate publication'.[79] It seems that an instruction of this sort did emerge from UNESCO, which later claimed (through Metraux) that there had been 'a misunderstanding'.[80] However, to read this dispute in such terms may be naïve. It seems more likely that Montagu and Metraux utilised the threat of early publication to cajole a reluctant new panel into agreeing on a final script, which was eventually published in July 1952. That such techniques may have been necessary reveals something of the struggle that was taking place at the heart of international scholarship between social and natural scientists over the meaning of race. At the core of this dispute lay the biologists' determination to hold on to race as a concept and not to close off the possibility of a future discovery of racial mental differences. Despite the best efforts of Montagu and Metraux, neither the first or second Statements could smooth out this ongoing conflict.

Ultimately, UNESCO did not display a united front of international science but instead highlighted a continuing scientific uncertainty about how to handle race as a concept. Rising political tensions fanned the flames of these disputes. By the mid 1950s, UNESCO's political agenda was no longer dominated by concerns about the repudiation of Nazism. In the grip of the Cold War, new agendas were in operation which ensured that the publication of race research remained politically charged, notably concerning the segregation issue in the US and apartheid in South Africa. The impact of these new political tensions on race research can be seen through the difficulties experienced by British biologist and educator Cyril Bibby in his dealings with UNESCO.[81]

Bibby was commissioned by Metraux in September 1953 to write a book about race and education which could serve as a guide to secondary school teachers across the world.[82] From the outset Bibby was concerned that such a book may upset the sensibilities of some UNESCO member states, notably the United States and South Africa. He warned Metraux:

> Although...I shall try to be really diplomatic, I do not see how an honest treatment can possibly avoid giving offence to some circles in USA, South Africa etc. May I take it that UNESCO is prepared to take this risk, providing that I avoid any <u>unnecessary</u> offence?[83]

Metraux's response reveals something of his inclination to drive antiracial study forward. He replied to Bibby: 'It is always good to be diplomatic, but remember the old proverb about Plato and truth: I prefer truth to diplomacy.'[84]

Suitably reassured, Bibby agreed to the UNESCO contract offered and had completed the first draft of the manuscript by April 1954. At this stage began a long process of delay, later recalled by Bibby as an attempt by UNESCO 'to "block" publication'.[85] Initially, UNESCO informed Bibby that it wanted his findings to be the subject of discussion in a special committee of member states prior to publication.[86] This had been part of the deal from the outset and did not in itself constitute any major indication of a policy shift.[87] Rather, it was a reflection of a newly developed sense of caution on the part of

Metraux, who seems to have been scarred by the level of criticism that surrounded the UNESCO race statements. He explained to Bibby:

> I have not forgotten how much UNESCO was taken to task by British scientists for having published the statement on race too hurriedly. The memory of this bitter experience has bred a caution which I think is a product of international civil service.[88]

The conference did demand some changes to Bibby's work, but endorsed its general content. Bibby was informed that he would be paid an additional $300 for making amendments and that the book would go to the publishers at the beginning of 1956.[89] Having made the changes, Bibby soon received the proofs of the final text and was told that the book would be out in a further month's time.[90] However, it did not actually appear for a further three years.[91] Bibby's book was instead held back by continuing objections from some member states of UNESCO, especially the US representatives.[92] The Americans' objections to the study were ostensibly trivial but seem to have been intended diplomatically to delay or indeed prevent the book's publication.[93] It seems that the US delegation felt that an authoritative UNESCO study of education and race might fan the flames of ongoing conflict over segregation in its Southern States, an issue which was particularly focused on education.

When President Truman established a special Presidential Committee on Civil Rights in 1946 it called for the 'elimination of segregation' in education and challenged the scientific sagacity of this kind of racial separation.[94] These findings added credence to an already mounting campaign. In a series of key courtroom battles between 1951 and 1954, the National Association for the Advancement of Coloured People (NAACP) led petitions against segregated education.[95] Ultimately, anti-racial science played a defining role in the Supreme Court's dismissal of the 'separate-but-equal' principle in education in the 1954 *Brown v. Board of Education of Topeka* case. Here the evidence of various social scientists, notably Kenneth B. Clark, was crucial in shaping the court's conclusion that educational segregation necessarily had a detrimental effect on black students and was thus unconstitutional.[96]

In the context of the important role being played by experts in the segregation debates as they continued, the American government seemingly considered that an authoritative UNESCO publication championing educational racial equality might be controversial, radical and unhelpful. Bibby's book as it finally emerged indeed carried such a message, going as far as to provide anti-racist strategies and advice to teachers working in segregated areas.[97] Bibby encouraged these teachers to stand up to prejudice and 'not allow social cowardice to masquerade as professional integrity'. He called for them 'to correct the errors and make good the deficiencies of any books' which gave inaccurate information on race and reminded those in Southern US States that despite the 'delicate' local situations, 'constitutional declarations provide a clear statement of the legal basis for non-discrimination'.[98]

Bibby was informed by Metraux in 1957 that he needed to change certain passages that could 'cause offence to Americans'.[99] He found it difficult to cooperate, complaining to Huxley that Metraux's requests 'betoken the same sort of intolerance which the book was supposed to oppose'.[100] He did, however, make changes, but none sufficient to quell the controversy surrounding the project. In the end, he believed that UNESCO would never agree to publish, and that the body who had commissioned the study five years previously was now 'trying to suppress the work' because of the Americans.[101] This belief finally led Bibby to seek publication elsewhere. Exercising what he believed was 'an author's fundamental right to have his work published' he released the book in 1959 with Heinemann.[102] UNESCO, after a long period of deliberation, agreed to the citing of their role in the development of the text; a decision only made, to Bibby's mind, in order to avoid 'a world-wide scandal'.[103]

Bibby's difficulties are interesting because they reveal something of the changing political pressures which governed the international study of race after the war. Whilst the initial UNESCO Statement and other bold gestures (like Truman's commissioning of a Committee on Civil Rights) can be seen as having been shaped by responses to Nazi racism, by the mid 1950s a largely different set of political issues drove racial politics, a change that seems evident in the decline of Bibby's UNESCO project. For British scientists of race, three international issues loomed large: the politics of segregation in America, the apartheid policy in South Africa and the increased Commonwealth (black) immigration to Britain. This chapter will now consider the evolution of post-war British scientific thinking on race in the context of these issues.

Progressives and conservatives in the 1950s: a case of reunification?

During his trials with UNESCO, Cyril Bibby exchanged numerous letters with Julian Huxley. The two men were well acquainted, primarily because Bibby was the leading biographer of Huxley's famous grandfather

T.H. Huxley, and Bibby perhaps hoped that Huxley's high standing in UNESCO could help pressure the organisation into publishing his book.[104] Indeed, Huxley did try to help. He wrote to Bibby in 1958, informing him that he had spoken to Metraux and Tom Marshall, UNESCO's Head of Social Sciences, on his behalf.[105] This support was not only based upon Huxley's personal warmth towards Bibby but also on Huxley's support for the project. He told Bibby in an earlier letter: 'I think your book is admirable, and I am very glad it is going to be published.'[106]

However, Huxley's support for *Race, Prejudice and Education* was not unequivocal. Like the first UNESCO Statement, Huxley felt that Bibby had gone too far on the subject of racial equality, even as he supported his political agenda. Criticising his friend's analysis Huxley warned him that he was making assertions that went beyond scientific knowledge. These criticisms addressed many areas of Bibby's argument but focused most notably on the idea of racial mental difference. Presenting a case that was consonant with the findings of both UNESCO Statements, Bibby had explained in his text that there was no existing scientific evidence of mental racial differences. He did leave the door open to future discovery, but argued that even if mental differences were at some time proved, this would not necessarily confirm the idea of racial hierarchies:

> ....if at some future time it were shown that there were in fact some ethnic differences of an intellectual or temperamental nature, this would not necessarily imply superiority or inferiority. People may differ in all sorts of ways, but difference is one thing and superiority is another.[107]

Bibby's caveat seems to reflect the biologists' caution at UNESCO but it was not enough for Huxley. He responded: 'Surely most intellectual differences would imply superiority and inferiority? One simply must not obscure this issue, which is of great importance for long-term eugenic reasons.'[108] This criticism was not rooted in any lack of support on Huxley's part for Bibby's anti-racist agenda, but was instead based in Huxley's ongoing uncertainty about the intellectual validity of non-racial science. He did not wish to stop Bibby from fighting the good fight, but wanted to ensure that his friend did not leave himself too open to scientific challenge. Huxley advised that Bibby should not counter the established racial slur that Negroes had smaller brains by making the case that there was 'no relation between the brain size and intelligence'. This, Huxley told him, 'was certainly not true when you look at the matter on the broad scale. You must safeguard yourself in some way.'

Huxley's criticism here epitomised the 1950s perspective of those interwar progressive biologists who had been fighting racist science for the previous 20 years. Once perceived as radicals, Huxley, Haldane, Hogben and the like were now comparatively conservative on the race issue, even as they all remained committed to the fight against racist politics. As was argued in the last chapter, the inability of these scientists to unreservedly abandon the idea of racial mental difference left them behind a new generation of scholars, who, mostly from the perspective of social science, were able and willing to do so. However, the issue here was not only about specialism. Bibby after all was himself a biologist. As the 1950s continued, the progressive interwar scientists also suffered from the unstoppable effects of time. Their once radical work was increasingly perceived as yesterday's news. The influence of texts such as *We Europeans* had been so great that its radical message had become the new mainstream, no longer the end of anti-racist enquiry but the new starting point. New radical conclusions went too far for the old progressives, who were married to a generation of science which had now passed.

Epitomised in the changing internal agenda at UNESCO, the key political battleground of anti-racial struggle had also changed. In the previous generation, Britain's progressive biologists had developed and published their ideas as a challenge to Hitler's racism, with its obsessive emphasis on the racial difference of Jews. By the 1950s, the main terrain of struggle in European and American racial politics concerned black–white relations, an issue which was itself planted firmly in the context of the Cold War. This brought forth a different response from the progressives, whose defence of black people against allegations of racial inferiority and difference had never attained the certainty which generated such robust defences of European Jewry in the war. Not only were these scholars unsure about black racial equality but they also did not want to accede to what they perceived as political (unscientific) Soviet egalitarian readings of science.

Hogben, Haldane and Huxley remained to differing degrees unconvinced that black people really were equal to white. For Hogben, the issue was a matter of uncertainty more than belief. Writing about the independence of Ghana in 1956, he admitted to reservations about the end of colonialism. Hogben described independence as 'a challenge of unprecedented significance', playing out 'in a scenario before a world audience'.[109] It is interesting that even Hogben was still willing to countenance the idea that inherent weaknesses might undermine the new country. He noted:

> What inborn endowments, if any, set limits to the contribution which hitherto handicapped communities may henceforth make to an era of potential plenty for all mankind is an issue which experiment alone can clarify.[110]

Huxley was more certain in his analysis. In his mind, Africa's population was not ready for self-government, a view which he expressed at an education conference in Natal in 1960.[111] Huxley told the conference that he understood the difficulties of South Africa as a 'white minority, surrounded by a less developed non-white majority'. Furthermore, he asserted that he did not expect the white minority to cede power in these circumstances. Describing Africans 'by and large' as 'uncivilised in comparison with the Whites', Huxley said that it was clear that they were 'not yet ripe for anything like political equality'.[112] Although it may not sound like it, Huxley's address was actually designed to challenge the policies of apartheid. Nonetheless, whilst he lectured that apartheid would fail, criticised the regime for training Africans for 'permanent inferiority' and called for 'the co-operation of all races' he did so from within a belief system which still saw black Africans as generally inferior.[113] As he told his audience: 'It is true that most Africans are in many ways like children, but this does not mean that their elders in civilisation should not try to educate them and help them grow up.'[114]

Huxley and Hogben maintained their respective views until the end of their careers. Whilst Hogben was much less willing to be drawn over the idea of black inferiority, they both retained beliefs about the existence of physical and psychological racial differences, as did J.B.S. Haldane. To this end, Huxley and Haldane made clear their objections to new leftist post-racial analyses which failed to take account of the scientific facts of racial difference. Haldane recorded in his lecture notes:

> There is a tendency in 'left wing' circles to underrate the importance of racial differences, that is to say innate differences between peoples in different geographical areas, whose ancestors have long lived in such areas, because the assertion of such differences has been used as an excuse for oppression and massacre, whether by ancient Jews, English slave-owners, high caste Hindus, Boer clergymen, or Nazi SS men.

There is no doubt that racial superiority as regards certain characters is a biological fact.[115]

Similarly, Huxley made a defence of the science of race in his 1962 Galton lecture. Citing H.J. Muller on the subject of racial difference, Huxley argued that it was 'theoretically inconceivable that...marked physical differences should not be accompanied by genetic differences in temperament and mental capacities, possibly of considerable extent'. Mirroring Haldane's remarks, Huxley dismissed anti-racial studies as 'new pseudo-scientific racial naivete'.[116]

Objecting to anti-racial work in this way went against the grain for both of these men. That they felt the need to speak out says much about their belief in the continuing importance of the study of race and their discomfort with emerging non-racial (often Eastern bloc) scientific analysis. Whilst they almost certainly did not intend it, their retrenchment on the issue brought plaudits from some strange places. After one New York lecture by Huxley in 1961 he received a letter of support from Reginald Gates, the first correspondence from his former colleague since their falling out over Huxley's anti-racism in *We Europeans* 27 years earlier. In the political climate of the Cold War, Gates clearly perceived that Huxley was now his ally, similarly opposed to leftist dismissals of race as a concept. Praising his lecture, Gates wrote to congratulate Huxley on his analysis and to 'say how fully... [he agreed]...with it'.[117]

The tempting conclusion would be that the progressives had come full circle on the race issue. That they had opposed race in the fight against Nazism but now mourned the loss of racial study and feared the Frankenstein's monster of non-racial science which they had done so much to create. In an analysis of this kind one could argue that Gates's letter signalled a reunion with Huxley and the others from the *Journal of Experimental Biology*, now that all these scholars were agreed on the importance of ensuring the continued rigorous scientific study of race in opposition to the post-UNESCO position. At the edges of such a reading there may even be some truth. Certainly, Gates and Huxley shared many beliefs on race in this period, as they always had done. However, the suggestion of a conservative/progressive reconciliation does not generally hold water. For while the progressives did by this stage certainly have major doubts about the extent of new scientific dismissals of race, they never wavered in their determination to fight racial oppression, or from the general belief that racial difference should not be overemphasised or translated into dubious political policy. As we shall go on to see, whilst they may have been similarly wary of the Soviet Union, Gates and his bedfellows sat firmly on the other side of key political race debates in the 1950s, supporting segregation and apartheid, while Huxley, Haldane and Hogben all remained politically committed to opposing these forms of race in power.

Faced with the apartheid regime which was cemented after Malan's 1948 election victory, the progressives (as they had done in the face of Nazism) sharpened their anti-racial position in order to oppose what they saw as racial oppression. As has already been noted, Hogben had left his research post in South Africa because of his disgust at proto-apartheid policies, and never hesitated to criticise a regime which he described as 'a sinister social set up'.[118] Huxley and Haldane shared this view, despite their uncertainty about black equality. Above and beyond any scientific reasoning, they found South African racial oppression unacceptable. Huxley argued:

> the policy of the South African government is a racist one, though they do not proclaim it as a principle or doctrine, as Hitler did. But racism, whether openly proclaimed or operating under a semantic disguise, is based on the same belief that inspired Hitler – the belief in the inherent and permanent superiority of some races, the inherent and permanent inferiority of others. I said *belief*; I should rather have said *superstition*, for our modern knowledge shows that this is not true.[119]

Ultimately, the progressives saw apartheid as unenlightened and morally abhorrent, even as they understood and occasionally agreed with some of the ideas that underpinned it. This political perspective, which mirrored progressive views on US segregation, kept a divide between these scientists and conservative biologists. Even concerning the UNESCO Statements, conservative and progressive positions were not the same. Progressives may have been concerned that the anti-racial content of the Statements had been taken too far but they generally supported the aims and agenda of the UNESCO racial education programme. Conservative racial scholars were hostile to the Statements on an entirely different scale.

Although the first UNESCO Statements on race were met by significant criticism from many quarters they did usher in, or at least represent, a new era of science where the study of race was regarded with heightened caution, suspicion, and sometimes downright hostility. Gates, Keith and other Anglo-American racial conservatives were deeply dissatisfied with this new state of affairs.[120] For them, the Statements represented a change of atmosphere in which it was increasingly difficult to operate. Gates outlined his views on the subject to Blacker at the Eugenics Society, explaining: 'The anti-race propaganda over here has reached the point where the existence of races is actually denied – a form of lunacy it seems to me.'[121]

The correspondence between Gates and Keith in this period reveals their deeply felt frustration and disaffection. Both men felt that Jewish conspiracy was driving science away from the true study of race. In one 1950 letter Keith told Gates:

> I am altogether with you about the Cuckoo race and now understand Hitler's attitude. They are certainly bridling the USA and riding that country so that it will be safe for Jews...My belief is anti-Semitism will out do the racial antagonism to inter-marriage with the blacks. I congratulate you on the vigour of your head and pen.[122]

As Keith's letter indicates, Gates agreed with his friend on this subject (as we saw in Gates's response to the petition against him at Howard). Amongst other things, Gates and Keith both blamed Jews in this period for their failure to get research published.

Gates's papers from the 1950s reveal that a range of prestigious scientific journals including the *American Journal of Physical Anthropology*, *Science* and the *Journal of the American Society of Human Genetics* repeatedly refused to publish his work.[123] The reasons for these rejections were varied. No doubt sometimes they were a reflection of Gates's growing pariah status. On other occasions, rejection seems to have been more linked to his now outdated scientific practice.[124] Whatever the reason, Gates did not accept rejections lightly, always seeing conspiracy lurking behind his failures. His letter (after initial rejection) to the editor of the *American Journal of Physical Anthropology* is typical of many written by Gates in this period. He was, he wrote, 'outraged' by the editorial decision. 'I have published some 320 papers in the scientific journals of at least eight different countries, but I have never been treated like this.'[125] After yet another failed attempt to appeal his rejection by the *Journal of the American Society of Human Genetics*, Gates informed the editor, Herluf Strandskov, that he would not accept that the decision was an academic one. He told Strandskov: 'My experience as an editor and writer is much greater than yours.'[126]

It was nearly always Jewish power that Gates saw as the non-academic reason behind the rejection decisions. In 1949 he wrote to the journal

Nature complaining about its negative review of *Pedigrees of Negro Families*. The journal's response to Gates's ranting was typical of many in the period. The editor told him: 'Your letter of April 1 was most offensive. The accusations it contains are so preposterous that I shall not trouble to deny them. I would, however, have liked the opportunity to publish your views on the Jews in the States.'[127]

Gates was far from alone in his thinking about Jewish racial conspiracy. Keith certainly agreed with him, claiming in his correspondence that Jews deliberately attempted to undermine the racial integrity of their nations by promoting 'miscegenation' and were responsible for preventing the publication of work by those scientists who were trying to stop them.[128] He wrote to Gates in 1950:

> The chosen race wants all races to mix – leaving it out free to exploit the amalgamated mass. The C[hosen] R[ace] takes more and more power here: it is in charge of our R[oyal] A[nthropological] I[nstitute] …at least strong enough to suppress a paper I wanted to publish…[129]

Aside from revealing these views about Jewish power, the correspondence between Gates and Keith in this period is interesting because it highlights the extent to which both men felt marginalised and isolated in the wake of UNESCO. Keith especially seems to have felt totally lost in the new environment. Now well into his eighties, he could not accept the decline of racial study in natural science. He wrote to Gates: 'Our social anthropologists talk a language I don't understand and anthropology – as Huxley knew it – is dead in Britain.'[130] Only months before his death, an 88-year-old Keith sent his friend a final exasperated letter:

> My colleagues in anatomy, with whom I used to stand quite high now consign my writings and advisings (*sic*) to the W[aste] P[aper] B[asket]. And it is some years since the Anthrop[ological] Inst[itute] let me know that I and all I stand for are on the 'shelf'.[131]

Keith's despondency was tempered by his belief that in time he and Gates would be proved right about race and about Jews. Although he knew that the day would not come soon enough for him, he expressed to Gates that it gave him 'great comfort' to know that he would continue to fight on.[132] This would, Keith was sure, allow for common sense to return to science. He explained to Gates: 'Your time and mine is not yet.'[133] For now, there was little to do except to continue to speak out against the grain of scientific opinion. On Gates's seventieth birthday, Keith sent a birthday message which summed up the relationship between the two men. He told Gates: 'you are a missionary of an uncommon kind. Good luck in your 70th year...so well spent. I damn your critics.'[134]

By the mid 1950s Keith was dead and Gates's mission was in peril. Although he had survived his trials at Howard and retained a research fellowship at Harvard, this contract was only temporary. As expiration approached, Gates could not find another post. Eventually E.A. Hooton, the Harvard anthropologist who had been instrumental in securing Gates's services in the first place, was charged with the unpleasant duty of gently reminding his friend that he was no longer entitled to use Harvard notepaper in his correspondence.[135] Largely because of his extreme racial views, Gates's isolation was becoming increasingly pronounced. In typically acrimonious editorial correspondence, the chairman of the editorial board of the US-based journal *Science* saw fit to dismiss Gates as 'a lone voice which, we are sure, is crying in the wilderness'.[136]

Gates was certainly on the back foot in this period but he was not inclined to give up lightly. Determined that he was right on race and that the matter was of supreme importance, he worked hard in the post-UNESCO environment to secure new alliances and outlets for the dissemination of his racial views. Gates's mission, as Keith described it, made him some important new friends, including a retired Scottish colonel and anthropologist, G.R. Gayre, who began corresponding with him in 1956.[137] Gayre initially wrote to Gates to tell him that he admired his 'great industry and enormous ability' and supported his view that, on the subject of race, 'the whole world is mad'.[138] In particular, Gayre shared Gates's feelings of isolation in the new post-war environment. He told Gates: 'There are few of us today who have continued to approach anthropology, and particularly ethnology, uninfluenced by the political pressures of our time, which have run to idiotic lengths.'[139] In another letter, Gayre congratulated Gates for resisting this pressure, assuring him that his work stood out 'as a monument of common sense and scientific objectivity'.[140]

Gayre also shared Gates's bitterness and anger, fuelled by similar beliefs that sinister political forces were at work in science, amounting in Gayre's mind to a 'deliberate conspiracy' to keep the truth 'from the people'.[141] Science was being attacked, Gayre wrote, by 'peddlers of lies, who are deliberately distorting facts to serve the ends of Leftist politics'.¹⁴² Like Gates, Gayre saw Jews at the centre of this conspiracy.

> I am afraid that Jews have got a grip on our Universities, as they had on Germany before the Nazis threw them out. For I hear Jewish voices coming constantly on our wireless, and all with jobs in British Universities, beating the big drum against race, and distorting the true facts of anthropology in order to do so.¹⁴³

In 1956, Gayre had been removed from his academic post at the University of Saugor in India as part of what he perceived as a post-independence drive to get rid of foreign academics. Like Gates, he was now struggling to find a new post and had no doubts as to why this was. He confided to Gates:

> ...you and I are not persona grata with those who dominate selection boards these days. We are labelled as 'racialists' and the whole political and ethical powers of the universities are to be organised to ban us, and our pernicious teachings.¹⁴⁴

Gayre's concerns seem to have been justified. Both he and Gates were labelled as 'racialists' and were unlikely to gain major academic posts as a result. In this context, they began to work together on a strategy that could rejuvenate racial study, and their careers, in this atmosphere. They knew that the political environment could attract allies as well as enemies, especially in the United States. Here, powerful American segregationists were increasingly on the lookout for scientists who could help them to justify their political agenda and rebuild the case for segregation, which had been shattered by the Brown ruling.

In a key example of the changing political issues which governed the international study of race in the 1950s, some scholars have argued that *Brown* in fact triggered a renaissance of scientific racism.¹⁴⁵ The effectiveness in this landmark case of anti-racial scientific analysis from Clark and others motivated pro-segregation forces to challenge the decision armed with their own scientific evidence instead of focusing, as they had done previously, on the principles of 'civil liberties' and 'state sovereignty'.¹⁴⁶ Newby has thus argued that the segregationists 'became convinced that social science was the Achilles heel of the Brown ruling...A challenge based on science...might succeed where those using states' rights and constitutional literalism had failed.'¹⁴⁷

In the *Stell* v. *Savannah-Chatham County Board of Education* case of 1963 and the *Evers* v. *Jackson Municipal Separate Schools District* case of 1964, the segregationists tried and failed to force a legal revision of the science behind the *Brown* ruling, using a range of testimony provided by sympathetic American academics.[148] The failure of these challenges did not end new relationships between politicians and scientists. On the contrary, faced with the growing Civil Rights movement and rising international hostility towards the idea of racial division, segregationists and their backers continued to utilise science in a range of ways in order to fight their corner.

This segregationist interest and money was a godsend to Gates and Gayre. As a true supporter of segregation now living in the US, Gates established contacts with various American pro-segregation organisations soon after the Second World War. In 1955 he joined the mailing list of the Jackson Citizens' Council, sharing their goals of resisting integration (especially intermarriage) and opposing 'pseudo scientific brainwashing' on racial equality.[149] Whilst he never actually became a card-carrying member of these kinds of groups, by 1958 Gates was receiving major funding support from wealthy individuals within the pro-segregation lobby who were well aware of the political benefits of his scientific output.[150]

Segregationist money ultimately paved the way for the establishment of a new journal in which Gayre would have responsibility as editor, Gates would serve as honorary associate-editor, and both would have a new way to disseminate their views and fight for the continuation of racial study in Western science. This journal was the *Mankind Quarterly*.[151] The third member of the *Quarterly*'s editorial team was the retired Columbia psychology professor Henry Garrett, one of the most frequently used expert witnesses of the pro-segregation lobby.[152] He joined Gates as the journal's other honorary associate editor.

The segregationist backing of the *Mankind Quarterly* was initially kept a secret as its editorial team attempted to present the journal as an objective scientific publication.[153] Gayre's statement of goals, in his first editorial, was entirely scientific. He claimed that the *Quarterly* had been founded to counter the 'strong leaning towards discarding heredity as a valid criterion in the study of man' and protect scientific ideas that, he claimed, were being 'deliberately suppressed'.[154] It was ostensibly for this reason that the *Quarterly* gave over so much space to discussion about race. As Gates put it: 'a journal should exist in which the problems of racial origin and racial relationships can be quietly discussed, without rancour or bigotry and with the primary aim of elucidating facts'.[155]

However, the *Mankind Quarterly*'s primary purpose was not the elucidation of facts as such. More accurately, it was designed to be a mouthpiece for the segregationist movement and a vehicle for the editors and their fellow travellers to strike out at their ever-growing number of professional enemies. As Gayre explained to Gates in one 1960 letter, the *Mankind Quarterly* 'will help things a good deal for all of us – as it means that there is something in which we can hit back'.[156] Gayre, Garrett and Gates's racial thinking meshed seamlessly in the *Quarterly*, which at once reflected both their political and scientific views. There was, though, a perceived need to try to hide the link between the two in order to present a façade of objectivity: As Gayre argued in a letter to Gates:

> While there is no doubt that a genetic and ethnological conception of race must invariably give support to the Southern States, and South Africa, at such a time as this it would be ill-advised to come out too openly. Consequently I am proposing that we should be a little 'Jesuitical' over the whole matter, and rather give ammunition to those who are trying to maintain sanity in the present colour hysteria which is going on in the democracies, rather than to enter the arena ourselves.[157]

In this context, the true backers of the *Quarterly* and its political goals were kept hidden.

The initial launch costs of the journal were met by a secret $2500 donation from a reclusive textile baron from New York, Wickliffe Draper.[158] Draper had a reputation as a leading pro-segregation donor who had funded, amongst other segregationist causes, the Mississippi State Sovereignty Commission's court battle against the Civil Rights Act at a personal cost of over 200,000 dollars.[159] Any open association with such a backer would have shown too much of the *Quarterly*'s hand, revealing its relationship to the pro-segregation lobby. Nonetheless, the journal's establishment and continuation were rooted solely in this kind of support. After Draper's initial informal handouts, long-term funding for the *Mankind Quarterly* was secured through the International Association for the Advancement of Ethnology and Eugenics (IAAEE), an organisation which had been created by Draper to support scientific racism in a variety of forms.[160] The relationship between the IAAEE and the *Mankind Quarterly* was so close that Tucker has seen fit to describe the journal as 'the house organ of the group'.[161]

It would be a mistake to argue that this funding drove the *Mankind Quarterly* to give a certain amount of coverage to the segregation issue. Whilst it is true that Gayre must have been constantly aware of who was sustaining what was an otherwise uneconomical publication, the promotion of racial segregation was a cause to which all of the *Quarterly*'s editorial team were as dedicated as their paymasters.[162] That, and the general promotion of racial study, was after all what the *Mankind Quarterly* had been created to do.

It was a goal that was not lost on other scientific communities, many of whom responded with hostility to the birth of the *Mankind Quarterly*. *Current Anthropology* led the charge against the journal, publishing an article by Mexican physical anthropologist and UNESCO contributor Juan Comas, which directly accused the *Quarterly* of being racist.[163] Empowered by the secrecy which shrouded the funding of his journal, Gayre was able to forcefully deny Comas's charges in a way which would not have been feasible had Draper's hand been visible. With some audacity, Gayre ranked hostility shown towards the *Mankind Quarterly* as similar to the attacks that had been made against the theory of evolution. Criticism of the *Quarterly*, Gayre informed his readers, had 'its nearest parallel in that which faced Charles Darwin from the entrenched forces of scientific and philosophical dogmatism and error in the last century'.[164] Determined to keep his cause in the best possible company, Gayre illustrated something of the aura of martyrdom that often accompanied racial conservatism in this period, by responding to the suggestion that his journal was only small with the reply: 'The *Mankind Quarterly* is a small journal – but so are the Gospels small books.'[165]

Forcefully proclaiming its scientific objectivity all the while, the *Mankind Quarterly* strove to preserve racial segregation. The journal set out its stall in its first edition, where an article by D. Purves argued that segregation was both natural and essential. The ideas in this article closely echoed the writing of the late Arthur Keith, who, Purves bemoaned, now 'received very little public attention'. It argued that race consciousness served a 'biological function' and represented 'instinctive feelings rooted in man's nature'. Focusing directly on contemporary affairs, the article went as far as concluding that altering the natural state of race separation could lead to disastrous repercussions:

> There is ample evidence to indicate that social instability is inevitable whenever attempts are made to create multi-racial communities, and a number of apparently insoluble situations already exist in Africa and the United States. In view of the instinctive character of the

antagonisms inherent in the multi-racial community, there appear to be good grounds for preventing the establishment of communities of this kind in the future and for attempting to solve the problems of existing communities by resolving them into separate societies on a racial basis.[166]

These kinds of arguments were made in virtually every early edition of the *Mankind Quarterly*, which became the main voice of pro-segregation science in this period, openly courting apartheid South Africa and defending American segregation.[167] The kind of conspiratorial anti-Semitism which governed Gates's world view was also evident.[168] That the *Quarterly* was not more obviously anti-Semitic should be seen as a matter of tactics, not belief, for the views of Gates, Gayre and Garrett on the subject of Jews were well documented. The reticence of the journal was perhaps rooted in an attempt not to pick one fight too many in what was an already controversial terrain of struggle and not to take any steps that would obviously link the *Quarterly* with Nazism.[169]

Given that the *Mankind Quarterly*'s main aim was to support the fight for Southern segregation, the centrality of two British scientists in the leadership of the journal requires some consideration. After all, the journal had much more to do with America than Britain, and indeed most of the contributions (as well as the funding) came from the US. Nonetheless, two-thirds of the *Quarterly*'s editorial team were British and the journal was based in Edinburgh until 1978.[170] To some extent, this can be seen as an accident of fate, although this does not provide a fully acceptable explanation.

Gayre set up the journal in Edinburgh seemingly because he retained publishing connections in the city.[171] There were few other reasons to base the *Mankind Quarterly* in the UK. The editors certainly did not think that Britain was going to be the intellectual base of the journal. Gayre was in fact exasperated by the lack of interest and warmth towards the *Quarterly* in British scientific circles. He complained to Gates:

> I have written to a considerable number of people in this country and I am getting no support...The fact of the matter is that Britain is really the centre of the whole of that scientific thought which runs counter to an interpretation of heredity in man.[172]

Gayre's understanding of scientific attitudes towards the *Quarterly* in Britain was shaped by his inability, despite serious effort, to bring other

British scholars onto the journal's editorial board. He told Gates: 'We now have quite a good Board, but excepting for you and myself we are the only British people! This is rather significant!'[173] This significance is doubled when one considers that neither Gates nor Gayre had worked in Britain in the previous 20 years. Seemingly, no university-affiliated academics in the UK were prepared to join the *Mankind Quarterly* team.

However, this lack of British presence need not necessarily be seen as rooted in unequivocal scientific disapproval towards all the work of the *Quarterly*. In the above analysis of the progressive end of British biology in this period, this study has argued that most British scholars were to varying degrees sympathetic towards the stated goal of the *Mankind Quarterly*, of challenging the tendency towards 'discarding heredity' which was becoming common in this period. This was, in fact, a goal which could have brought together most of Britain's biologists in the 1950s and 1960s. The isolation of the *Quarterly* can therefore only really be understood through a political/ideological lens.

Whilst it may have kept its funding secret, the journal's political aims (pro-segregation and pro-apartheid) were clear for all to see. Even if other scientists agreed with these aims, and for the most part they did not, joining the editorial board of such a journal would have been a strong statement for any affiliated British academic to make. There may well have been some truth to Gayre's assertion that many of those who he had expected to join the journal suffered at the last moment from 'cold feet'.[174] For one thing, as we shall see below, the *Mankind Quarterly* was not the only scientific organ in Britain to receive funding from Draper in this period. It was however the most overt manifestation of racial research, in an era when few scientists were prepared to 'go public' about their ongoing belief in racial study.

Buoyed by its American financial backers, the *Mankind Quarterly* managed to survive despite the reluctance of most scientists to support it in any way. Whilst at some levels it can be dismissed as marginal, it is perhaps fair to say that at least in terms of its scientific goals it represented more of British scientific opinion than the absence on the editorial board might suggest. Instead, the refusal of British academics to take part in the enterprise should be seen as an indication of the ongoing political divisions between progressives and conservatives on the issue of race. Whilst progressives like Haldane, Huxley and Hogben may well have agreed with some of the scientific principles of Gates and Gayre, they found the political translation of these ideas, as represented in the *Mankind Quarterly*, repressive and abhorrent. Haldane's 1963 review of Weyl and Possony's *The Geography of Intellect* illustrates the point.

As we have seen, Haldane retained an ongoing belief in racial mental difference in this period, yet he unleashed furious criticism on these regular contributors to the *Quarterly*. Under the title, 'A Text for American Fascists', Haldane outlined what he perceived to be the possible consequences of texts of this kind and the constituency to which he thought the book would appeal.

> The authors are not, I think, fascists. But similar books by Germans who were not adherents of the NSDAP provided Hitler with all the intellectual ammunition which he needed for his racial policies. An American Hitler is certainly not an impossibility...This book, will, I do not doubt, satisfy a widespread demand, and is quite likely to become a best seller, like the more widely advertised pulmonary carcinogens.[175]

Haldane's criticisms of the science in the text were more moderate. Oozing caution, Haldane did not wholly dispute *The Geography of Intellect*'s case concerning mental racial difference but only commented that 'most geneticists' had rejected the idea of 'large genetically determined differences between the mental and moral capacities of different human races'.[176] Ultimately, Haldane's challenge was political, not scientific. Geneticists, he wrote, in his notes:

> ...must consider their social function, which differs in respects from that of other scientists. The practical applications of genetics, such as the production of hybrid maize, have perhaps been uniformly desirable. But the influence of genetic ideas, such as that of racial purity, has often been disastrous. We have the duty to spread what we believe to be the truth, but it is extremely hard to avoid exaggerating the genetical evidence which we may think favours what, on non-genetical grounds, we regard as ethically or politically desirable aims.[177]

It was for these reasons that Haldane took issue with *The Geography of Intellect* and for these reasons that the progressives remained in this period divided from the likes of Gayre and Gates over the issue of race.

This chapter noted earlier that Britain's leading progressives had been on the retreat from the political soapbox ever since the Lysenko affair. This retreat partly explains their reluctance to subscribe to UNESCO's populist position, especially as the first UNESCO Statement would have been seen to some extent as a pro-Soviet scientific gesture.

It also perhaps explains their reluctance to join the *Mankind Quarterly*. They may have felt that science was making a mistake in its too hasty retreat from race. In some cases, this meant that leading progressives (like Huxley) continued to play a role in moderate if conservative movements like the Eugenics Society. However, progressives shuddered at the thought of being a part of a scientific movement which openly gave succour to political racism like the *Mankind Quarterly* did. Long after they had finished drawing any meaningful critiques of the use of race in science, and long after they had lost interest in trying to lead British social and political discourse, the progressives still wanted to sit on one side of the good fight.

The progressive/conservative fault line in British science therefore remained, as the progressives put aside their own doubts and beliefs about race and refused to publicly speak out in favour of racial study. As Haldane noted, 'any arguments which refute such beliefs are a contribution to the cause of peace'.[178] The progressive trump card in this battle was the lack of scientific certainty on race. Whilst they mostly themselves believed in racial difference they knew full well that it could not easily be proved. Armed with this absence of evidence they could resist both those who dogmatically asserted difference and those who called for them to subscribe to the egalitarianism of UNESCO. Haldane knew as much, commenting on this strategy: 'In stating our ignorance we shall also be stating the ignorance of those who claim to know the answers.'[179]

British science and post-war Commonwealth immigration to Britain

In the context of British society, by far the most pressing racial issue in the post-war period concerned the arrival and settlement of black Commonwealth immigrants. Black people had always lived in Britain but the combination of a labour shortage, poverty in South Asia, Africa and the Caribbean, ever-improving transport links and Empire-inspired identification with Britain drew unprecedented numbers in the 1950s.[180] Although many within British society were uncomfortable to varying degrees with the increased black presence, there were no effective legal means in place to restrict this immigration. In fact, the 1948 Nationality Act had entrenched the right of any subject of the British Crown to live in Britain.[181] This legislation neatly expressed Britain's self-perception and desired world role as Commonwealth 'mother country' but it was not intended to actually enable significant migration to the UK from the black Commonwealth. Amid the symbolic grandeur of the Act, the full potential of its immigration implications was not, it seems, considered.[182]

Despite the unforeseen consequences, the government was keen to avoid rescinding or amending British immigration law, as regards the Commonwealth, for several reasons. There was a fear that restrictive legislation might provoke an angry reaction from the black Commonwealth countries. Changing the nationality legislation, it was thought, may lead to hostile interpretations of Britain's outlook. As one official put it in November 1961, it would be perceived that 'as soon as people of the wrong colour started coming through the door in substantial numbers, we took steps to close it'.[183] But perhaps more importantly, the government did not want to take any action which could be labelled as racist. In the post-war, post-UNESCO atmosphere, racial immigration legislation would have undermined the liberal self-perception of the nation and aligned Britain with unwanted bedfellows like the apartheid regime and Southern segregationists in the US.[184] However, in 1962 the government did enact a law to restrict the entry of Commonwealth immigrants to Britain in the form of the Commonwealth Immigrants Act. This chapter will explore the views of British race experts on the issue of black immigration and the possible need for restriction and will assess where the government stood in relation to these expert opinions. Hampshire has argued that the 1950s and 1960s witnessed 'an exponential growth in "expert" interest in race relations and immigration'.[185] Indeed, this was a period where a glut of specialist texts and reports were written on aspects of immigration and its implications for British society. This chapter will explore this scholarly interest and assess its impact on wider social responses to the heightened black presence in Britain.

Commonwealth immigration, as Stone has highlighted, seems to have sparked concern and alarm in the Eugenics Society.[186] That it did so indicates that beneath the Society's carefully nurtured image of an organisation concerned solely with population issues remained a community still very much interested in questions of racial difference. In 1958 the Eugenics Society showed something of its hand on the issue, through the publication of a pamphlet by the organisation's General Secretary G.C.L. Bertram.

In this pamphlet Bertram made a case against Commonwealth immigration which was rooted in objections to the UNESCO position on race. Like other critics in Britain, Bertram argued that the Statements' declarations of racial equality, and their calls for the end of the use of the term, were based upon political agendas and not scientific fact. Bertram explained that whilst he shared the laudable sentiments of the UNESCO position, he would continue to use the word in recognition that it served an important function, explaining what he considered to be inherent human differences.

> The use of the word 'race' in this Broadsheet is unavoidable. This term, for groups of human kind with obvious genetic differences, is at present unfashionable. The United Nations Organisation and its derivatives (UNESCO, etc.) has been largely responsible. A benevolent spirit has animated the participants in non-political conferences and documents: stress has been laid on the admirable potentialities of the varied groups of mankind. However, actuated no doubt partly by political necessity, and in order to stress similarities and the propriety of equality of opportunity for each child born into the world, the quite obvious dissimilarities between people and individuals have been minimized.[187]

Bertram went on to argue that further West Indian immigration to Britain was undesirable because of the potentially damaging effects of racial mixing on the national stock. In what was an inversion of the progressives' usual defence of racial mixing, Bertram did not cement his case with clear evidence but instead highlighted the absence of such information. He conceded that hard facts which opposed mixing were not forthcoming, but argued that it should be discouraged in the name of caution. He pointed out that 'miscegenation' was 'irreversible', 'that it is easy to mix and impossible to unmix'.[188] Bertram's case neatly expressed the dominant sentiments of the post-war Eugenics Society. At the core of the pamphlet's argument was the idea that races did exist and that racial mental difference was a likely truth.[189] These views, as we have seen, were hardly controversial in British science and mirrored the progressive biologists' position on race in this period. With a moderation which had been dominant in the Eugenics Society since Blacker's reign, the pamphlet firmly rejected the idea of any measure of force to prevent racial mixing, arguing that even if such action were 'desirable in principle', it would 'require... a removal of liberty which should not be tolerated'.[190] Reminiscent of Huxley's 1936 Galton lecture, Bertram's argument conceded the role of social factors, alongside issues of heredity, in shaping popular concerns about 'miscegenation'.[191] All in all, if the Society's decision to speak out on this issue showed something of its ongoing interest in race, it also showed something of the kind of moderation which had dominated British eugenics since the 1930s. With this tone and outlook, the Society could and did attract mainstream scientific members and head off most of its adversaries. It is significant in this context that Julian Huxley felt able to assume the Presidency of the Society in 1959.

However, there was another side to the Eugenics Society's interest in racial mixing in post-war Britain, which brought it as close to the US segregationist camp as it was to the British progressives. It is revealing of the tightrope that the Society was walking in this period that the edges of this other perspective were also visible in Bertram's pamphlet on West Indian immigration. Here Bertram made a case, reminiscent of Arthur Keith, that racial prejudice was not only natural but part and parcel of the development of the human species. Races, he argued, had naturally developed in relative isolation until technological innovation increased the potential for human movement 'on account of modern transport'. These new possibilities, Bertram continued, were 'not necessarily a good reason for a switch in the direction and pattern of human evolution' and had the potential to cause 'emotional and genetic chaos'.[192] Like Reginald Gates, Bertram contended that there was little reason to believe that the mixed race populations that would occur as a result of more immigration would be of a decent racial quality. He argued: 'few will assert that there are, in our own species, obvious examples of a mixed population which is all the better for being the progeny of two markedly dissimilar parent stocks'.[193]

These kinds of arguments brought the Eugenics Society to the attention of Wickliffe Draper, the benefactor who had funded the *Mankind Quarterly* and other American pro-segregation scientific organisations in this period. Draper's representative Harry Weyher met with Bertram in February 1961 to discuss the possibility of setting up a British institute to study race, paid for by Draper.[194] Such an institute was not in the end created but Draper did agree to fund two research fellowships (worth 1000 dollars each) to be held under the auspices of the Eugenics Society.[195] Reginald Gates played a pivotal role in these negotiations.[196] Draper clearly trusted him and wanted him to play a central part in managing his donations to British projects. Weyher's correspondence reveals that Draper wanted Bertram to discuss suitable projects with Gates and was inclined to give more money if projects were directly supervised by Gates or Bertram himself.[197] Indeed, the fellowships were not offered until Gates had written to Draper expressing his support for the idea.[198]

Despite Gates's actions as a middleman, the relationship between the Eugenics Society and Draper did not begin smoothly. The Society rebuffed Draper's initial offer of funding because it felt constrained by his stipulation that he should personally approve the subject of research undertaken in the fellowships. The Eugenics Society would not accede to this demand, seemingly because they felt that it undermined their scholarly independence and integrity. Bertram informed Weyher that the Society believed that the submission 'of precise individuals and their research projects to your client [Draper] for acceptance or rejection is out of accord with most modern practice and would be likely to be unworkable'.[199] This was a rebuttal not born from malice or lack of interest but instead from the financial confidence of a Society which, since Henry Twitchen's interwar bequest, felt able to deal on its own terms. This indeed proved to be the case. Draper retracted his insistence that he must approve projects personally, and instead gave Bertram executive power to select the projects himself.[200] This compromise was good enough for the Eugenics Society who now felt able to take money from their segregationist backer. Bertram commissioned a study of race mixture in Liverpool to be conducted by the anatomist G. Ainsworth Harrison and an American approved by Draper named John B. Trevor. The other fellowship was given to Prof. Sargant Florence at the West Midlands Social and Political Research Unit in Birmingham to explore the fertility of immigrants.[201]

Whilst neither of the project reports carried the kind of racist conclusions that would have greatly pleased Draper it is nonetheless significant that the Eugenics Society was prepared to have even a surreptitious relationship with a man described by Gates as 'strongly anti-Negro'.[202] Taking Draper's money could be seen as a pragmatic decision which enabled the Society to drive forward its research agenda without making any obvious concessions to Draper's world view. However, it could also be seen as a reflection, albeit to a limited extent, of a shared terrain of belief and interest. Like Draper, the Eugenics Society was opposed to further racial mixing and inclined to try to discourage it through education and research. Whilst the Eugenics Society's focus was on black immigration to Britain where Draper's was on the maintenance of segregation, there was an obvious similarity of agenda. If the leaders of the Eugenics Society viewed Draper's extreme politics with distaste, it was only enough for them not to sanction his directing of their organisation, not sufficient to decline his donations. Ultimately, whilst they were more moderate and more diplomatic, Britain's Eugenics Society had more than a little in common with Wickliffe Draper. Most importantly, they generally shared the idea that the growth of a multiracial society was not to be encouraged.

The majority of experts on black Commonwealth immigration in 1950s Britain did not share these beliefs. These scholars, mostly social anthropologists and sociologists, were generally more sympathetic to the presence of Britain's new black population than were the eugenicists.[203] There were two primary reasons for this difference in outlook. Firstly, academics that had been trained in social scientific disciplines were generally less interested than natural scientists in issues of stock and racial regeneration.[204] Their work tended to focus on immigration in terms of its impact on labour, housing and social interaction, leading to concerns which were significantly different from the biologists' fascination with 'miscegenation'.[205] Secondly, the research of social anthropologists and sociologists developed from a politically different standpoint. In the main, these researchers were politically sympathetic to black newcomers. In many ways, the tools of their trade had been cut in the US, where leading experts had developed research on race as a conscious part of a political anti-racist struggle.[206] The role models of the emerging social studies community in Britain were scholars like Gunnar Myrdal, Otto Klineberg, Kenneth B. Clark and Ruth Benedict whose work, though they may have claimed it as objective, had played and was continuing to play a vital part in the ongoing political battle against segregation.[207] Following to any extent the scholarly approach of these people necessarily ensured a response to the growing British black community which was significantly different from that of physical anthropologists, eugenicists and biologists.

The pioneering social study of the black population in Britain was published by Edinburgh anthropologist Kenneth Little in 1947.[208] Little's influence lay not only in his own work but in his training of some of the most important of Britain's future experts in race relations including Michael Banton and Sheila Patterson, scholars who developed something of a collective identity as part of the Edinburgh University Department of Social Anthropology.[209] At the same time that Little was conducting his research in Cardiff, the social anthropologist Anthony Richmond was studying the impact of black immigration in Liverpool.[210] Both men were influenced by American studies of race and both were inclined to be sympathetic towards black immigrants.[211] In his preface to *Negroes in Britain*, Little showed his hand from the outset, informing his readers: '...my attitude is affected by sympathy for the victims of this prejudice, and by a considerable amount of irritation with the ideas and factors which underlie it'.[212] Little perceived the controversy that surrounded black immigration as a matter of temporary prejudice, not an issue of real human difference. Writing in 1950 he asserted: 'To future generations it may seem extraordinary and unbelievable that a slight difference in the chemical composition of their skin should have caused men to hate, despise, revile and persecute each other.'[213] Similarly, Anthony Richmond, in *Colour Prejudice in Britain*, saw fit to dismiss from the outset the possibility of black racial inferiority and the notion that racial feeling was rooted in any kind of natural phenomenon. Like Little, and like their American forerunners, Richmond believed instead that race needed to be understood as 'an attitudinal phenomenon'. Explaining the basic terms of his investigation he recorded:

> ...any interpretation of West Indian behaviour which is based on any assumption of inherited inferiority of mental capacities among the Negro peoples, or the children of mixed race parentage, was rejected...The author does not accept any explanation of racial antipathy which is based upon concepts of inherited instincts.[214]

The sociology scholars that followed Little and Richmond were mostly even clearer in their feelings about race as an idea. Ruth Glass saw fit to preface her study of the black population of London by noting that just as it was not necessary for criminologists to declare at the outset of research that they did not support murder, so she did not intend to state her abhorrence of racism at the beginning of her study.[215] In Glass's mind, it seems, opposition to racism could now be assumed amongst British social scientists. Describing the field of scholarship in a 1968 lecture, Philip Mason, sociologist and one-time Director of the Institute of Race Relations, described his peers as a group who were driven by the desire 'to reduce prejudice and improve relations'. Most scholars, he observed, were 'consciously involved' in trying to shape the racial views of the society around them.[216]

In this context, it is not perhaps surprising that the post-war scholars of British race relations produced a fare that was generally different from British biologists in terms of their conclusions about black presence in the population. Britain's new generation of social scientists were comparatively untroubled by the effects of immigration, describing and analysing social interaction generally without reference to biological ideas about the racial qualities of black newcomers. Instead of arguing that differences between immigrants and hosts were essential or natural, the sociologists' research frequently asserted that prejudice and discrimination were socially constructed phenomena which could and should be challenged through education.

As tensions rose about immigration in the late 1950s, several of these scholars attacked the government for not doing enough to educate public opinion about the new immigrants.[217] In Glass's study of black *Newcomers*, she argued that the government had failed to provide education for the public about the new immigrants, inaction which had led to disadvantageous, fantasised images coming to the fore: 'So long as the screen of information behind the coloured man is blank, he stands out strikingly as the dark mysterious stranger.'[218] Analysts commissioned by the Jamaican government to assess the immigration situation in Britain concluded similarly: 'The introduction of foreign workers into Britain was accomplished by the government without much preparation of the native population.'[219] In the conclusion of her study, Glass emphasised the need for an increased governmental role in immigrant reception and her belief that the present administration was failing to fulfil its responsibilities in this respect.

> It is the responsibility of Parliament to give tolerance a push; and to provide a code of standards on matters of race relations. Leaders of opinion will have to give the lead more emphatically than they have done so far.[220]

There seems to have been some consensus amongst social scientists about the measures which needed to be taken by the government to assist black people in Britain. Politicians were often criticised for not providing sufficient facilities and support to smooth black assimilation into British society.[221] Scholars suggested not only educational campaigns, but also anti-discrimination legislation.[222] Michael Banton made a forceful case for such measures in *The Coloured Quarter* in 1955. Whilst Banton conceded that anti-discrimination laws would be impossible to enforce in every instance, he argued that they would have 'the unquestionably great effect' of 'providing tangible evidence for the often voiced claim that Britain does not believe in colour bars'.[223] Along with new law, Banton called for improved anti-racist education, which, he argued, 'would do well to be based upon the material which has been provided by UNESCO in its campaign to improve inter-group relations'.[224] A similar case was made by Sheila Patterson in *Dark Strangers*, where she also called both for governmental action on public education and legislation to prevent discrimination. What was needed, Patterson argued, was 'various forms of nation-wide social action'.[225]

Other analysts of the immigration went further than these sociologists, accusing the government of deliberately neglecting black immigrants in an attempt to create sufficient hostility to justify restrictive legislation. In an emotive analysis, the journalist Paul Foot argued that the government had left underfunded local and voluntary bodies to absorb and assist the new immigrants in order to deliberately exacerbate racial tensions.

> While these organisations tackled, unaided, problems with which they were not equipped to deal, the Conservative Party prepared to reap political gain from the resentment and squalor which their own neglect had created.[226]

Foot's case was polemical, but the government clearly did not 'do enough' in the eyes of many scholars to smooth the path into the UK for Commonwealth immigrants, instead legislating to restrict their entry in the Commonwealth Immigrants Act of 1962. For many, the political reaction to racist rioting in Nottingham and Notting Hill Gate in 1958 epitomised government behaviour in the period. In their analyses of these riots, many of Britain's leading social scientists argued that the government had used the violence to catalyse the path towards Commonwealth immigration restriction, and not to do what was needed in terms of educating the public and/or entrenching anti-racist legislation. Banton concluded:

> The government of the day took no active steps to assist with problems in industrial cities which, in an era of very full employment, attracted the coloured job seekers. Local pressures for the restriction of immigration therefore mounted, and the Commonwealth Immigrants Act was introduced.[227]

Glass similarly contended, regarding the rioting, that 'a firm policy for racial equality in Britain' would have been sufficient 'to break into the vicious circle of tension between white and coloured at all levels and in all spheres'.[228]

In the end, the British government was seemingly pulled in two distinct directions by experts on race in this period. On one hand, many eugenicists and biologists argued that black immigration needed to be viewed with varying degrees of caution on the grounds that there was insufficient knowledge about the effect that it may have on the future of British racial stock. Often lurking within these arguments was the idea that hostility towards racial mixing (and assimilation in general) was natural and that black people were inherently different, probably inferior. At the same time as these arguments were being made, Britain's leading voices in social science were championing the first UNESCO Statement position on race as a social construct, not a biological reality. Instead of calling for restriction, these scholars mostly promoted the idea of education and legislation to smooth integration. Unlike the eugenicists, they seldom focused on the idea that immigration needed to be resisted for biological reasons (even if they were sometimes unconvinced that immigrants were the same in their values and make-up as indigenous Britons).[229] All the while Britain's leading progressive biologists, so troubled by political racism but unready to join the sociological abandonment of the biological concept of race, mostly kept their own counsel, unwilling to try to lead the government as they had done so forcefully before and during the Second World War.

Looking at the origins of the Commonwealth Immigrants Act it is evident that the government, whilst not explicitly influenced by either the eugenic or social science camp, retained a view on immigration which was more in tune with eugenicists. This may seem strange given that the Eugenics Society in the post-war period was, as Hampshire notes, 'a marginal and increasingly esoteric movement'.[230] The explanation perhaps lies in the distance between belief and political expression. The Eugenics movement in this period was marginal not because people did not agree with many of its ideas but more because, in the wake of Nazism, it was politically unfashionable. Thus Hampshire tells us: 'what is interesting is that the views of Bertram were not so dissimilar from the ideas circulating in the popular publications'.[231]

As has been noted earlier in this book, scientific ideas about race and immigration were differently constructed from political and social ones. However, as was the case in the 1930s and 1940s, a mirroring of scientific thinking on race and immigration is partially evident in the politics of the post-war period. The ideas that were mirrored were mostly those of biologists and eugenicists and not social scientists. As Britain faced increasing Commonwealth (black) immigration, concerns about the racial effects of mixing and changes to the racial composition of the nation made a significant impact on the case for restriction. As in science, these concerns were mostly expressed quietly, shrouded in post-Nazi and post-UNESCO awareness that such thinking was not socially acceptable.

In *Newcomers*, Glass not only recorded the prevalence of 'miscegenation' fears among white Britons but also commented that the level of these fears was as high 'as might be expected'.[232] To Glass and others who worked in this period surrounded by prevalent discussions of racial mixing and stock, popular hostility to 'miscegenation' came as no surprise. Other analysts have confirmed the influence of these ideas on post-war race relations. For example, Senior and Manley highlighted the strength of feeling about racial mixing in their survey of employment prospects for black people in Britain, recording that:

> White workers were much more likely to object to the hiring of coloured workers in situations where they might be thrown into contact with white women, than where the jobs involve only males.[233]

A 1951 conference of the British Council of Churches, addressing the issue of black immigration, similarly argued that it was the fear of 'miscegenation' (and not employment competition) which formed the main barrier to achieving black assimilation in Britain. The conference agreed a motion that 'the consequences of social mixing, even if it involves marriage, must be accepted if there is to be genuine assimilation'.[234]

Concerns about racial mixing also existed within governmental circles where, Hampshire has argued, 'there can be little doubt that interracial sexual relations were considered, a priori, to be undesirable'.[235] Analysis of a 1956 report by a Committee of Ministers reveals the extent to which anxieties about 'miscegenation' were prominent within government discussions of immigration policy. The report (which considered the possibility of immigration restrictions) argued: 'On present evidence a trend towards miscegenation can neither be forecast nor excluded. If such a trend were to occur it would be an important factor...[which could lead to]...broader problems of social assimilation.'[236] The idea that immigration may lead to undesirable mixed race relationships was again suggested in a government Working Party report the following year.

> In most districts where there is a coloured community of a few years standing, many coloured men are married to or living with white women of low social standing or low morals. The number of half-caste children in these districts is increasing and many are thought to be illegitimate.[237]

In this period, government statements that overtly condemned the idea of mixed race relationships were rare. Fundamentally, though, there is little reason to consider that opinion in Westminster and Whitehall differed significantly from the general public. Across Britain, fears about 'miscegenation' were commonplace and often accompanied by beliefs concerning black sexual deviance, debauchery and promiscuity.

Black immigrants were widely perceived as being responsible for a disproportionate amount of sexual crime.[238] Glass's report highlighted the prevalence of this belief: 'The stereotype of the coloured man, who induces white girls to become prostitutes and who lives on their earnings, was frequently mentioned.'[239] The sexual behaviour of black men became the focus of much anti-immigrant hostility during the 1950s. Analysis of the 1958 Notting Hill riots has highlighted that the main demand of white vigilante street squads was the deportation of black men who had been convicted of sexual offences.[240] Correspondence sent to the Parliamentary Secretary to the Minister of Education, Edward Boyle, by some of his Handsworth constituents, was typical of much of the popular hype surrounding the image of the black sexual criminal in this period. One hysterical letter included a newspaper extract and described a black man 'attempting apparently to coerce a citizen to consort with a woman, and even attacking the man'.[241] Whilst Boyle was troubled and irritated by these letters, it is clear that this image of the black man as a sexual criminal also permeated some areas of government thinking and decision making.

As early as 1953, the Working Party on Coloured People Seeking Employment in the UK reported the 'marked number of convictions of coloured men for living on the immoral earnings of white women'. These crimes, the committee minutes recorded, were 'far more widespread than the few prosecutions indicate'.[242] Four years later, the same committee that had warned of the dangers of racial mixing asserted that the Indian and Pakistani immigrants under investigation displayed 'a tendency to engage in brothel keeping and living off the immoral earnings of women'.[243] In the following year, a report prepared for a governmental response to an adjournment motion repeated this allegation, that 'certain types of immigrant possess a propensity to live on immoral earnings of women'.[244]

Beliefs in differing black and white sexual values permeated some people's thinking at every level of British society. Commonly central within these ideas was the notion that black male immigrants were more sexually liberated, attractive and promiscuous than white British men.[245] Belief in the 'super-sexual' black male ran deep within the mind frames of British people. Many of those who would have described themselves as having no racial prejudices still believed black sexual difference to be a matter of obvious fact. For example, Paul Foot outlined that venereal disease was more common among the black community due to 'their free and easy attitude towards sex' and their being 'less inhibited about prostitutes than the indigenous population'.[246] Likewise, Anthony Richmond argued that black men had trouble settling in Britain due to their belief in 'promiscuous sexual relations...as normal behaviour'.[247] Richmond concluded that 'sexual maladjustment is sometimes at the bottom of the criminal behaviour found among some West Indians'.[248] Sheila Patterson concurred in her analysis of the 'new' black community, *Dark Strangers*: 'Real differences emerged particularly strongly in the matter of values and norms associated with sex and family life.'[249]

If even liberal academic and generally pro-immigrant theorists like Richmond and Patterson did not always challenge ideas of black racial super-sexuality in this period, it is perhaps unsurprising that these beliefs were common currency amongst the general public and policy makers. Glass recorded:

> There is no doubt that a tangle of sexual images about the coloured man, and of sexual competition with him, strongly affects attitudes to coloured people, in general, outside the place of work.[250]

Despite his own ambivalence on the subject, Richmond too noted an obsessive public interest in black sexual difference. In a 1950 analysis of stereotypes, he cited the existence of 'strange beliefs about Negro sexuality, his high fertility and promiscuity' along with 'the belief that once a woman has had sexual relations with a Negro she will never return to a white man'.[251]

Views of this kind seemingly permeated government thinking on black immigrants. Anthony Eden, in a paper for Cabinet in 1955, echoed a belief in black promiscuity. Describing the black family, he argued:

> In some cases the woman does not even know the name of the father, who frequently contributes nothing to the support of his children and probably scarcely knows of their existence.[252]

The idea that black people were especially prone to dysfunctional family life was not only an individual view but permeated the outlook of government agencies with direct responsibilities for the new immigrants. A report from the National Assistance Board in May 1958 contended that the majority of black people were 'living communally, in the fullest sense of the term, a form of living which is their natural way of life'.[253]

Allegations of difference surrounding the act of sex and procreation should perhaps be read as a manifestation of deeper British concerns about the implications of interracial sexual contact; ultimately as a fear that 'blackness', through mixing, would alter the racial make-up of the nation. Cabinet minutes from 1955 highlighted this feeling, expressing concern that further black immigration might bring 'a significant change in the racial character of the English people'.[254] A Committee of Ministers report, the following year, reiterated the point. 'We clearly cannot undertake to absorb in such a densely populated island inhabited by a different racial strain all the coloured immigrants who may wish to come here.'[255] The colour of the immigrants, and the character traits perceived through this colour, was often seen as an insurmountable barrier to assimilation.[256] To many, 'blackness' was not and could not be British. Sivanandan has argued, regarding perceptions in the period, 'that a Black citizen was not completely a British citizen when he was a Black British citizen'.[257] Glass similarly noted:

> Although the coloured migrant is a British citizen by law, it is still true that he is set apart in this society – officially and privately; subtly or crudely; positively or negatively.[258]

The deliberate exemption of the Irish from the restrictions of the 1962 Commonwealth Immigrants Act as it emerged seems to support the idea that it was 'blackness' which was perceived as inherently foreign, 'alien' and undesirable.[259] Dummett and Dummett have argued, regarding the Act that:

> ...anyone genuinely concerned about total numbers, irrespective of colour, would be bound to seek control of the largest section of immigrants, the Irish, and, when a ceiling was imposed on Commonwealth immigration, to demand a similar ceiling for European immigration.[260]

There is indeed evidence that the British government's decision to exclude the Irish from immigration legislation was at least partially a racial one. The Irish needed separate consideration, G. Lloyd George argued in a 1955 note, as 'the Irish are not – whether they like it or not – a different race from the ordinary inhabitants of Great Britain'.[261] This view was not shared by everyone. Indeed many scholars have noted that Irish immigrants have historically often been racialised separately from perceived Anglo-Saxons within British racial discourses.[262] However, in this period, the Irish were seen by most policy makers as acceptable and assimilable immigrants who would not undermine British character and could be trusted as insiders within the fence of British racial preservation.[263] In this context, the Irish government was approached by the British administration, not to request the prevention of movement of their own citizens (as was the case with black Commonwealth governments), but to join Britain in her attempt to exclude other (black) immigrants.[264]

> ...the Home Office expect to be able to make arrangements with the Republican authorities whereby such people (and especially coloured immigrants) would not be allowed to enter the Republic except on terms similar to those prescribed by our proposed bill. It is thus hoped that Britain and Ireland will form what would amount to a continuous immigration ring fence, as is at present the case in respect of aliens.[265]

This arrangement identified black immigrants, in contrast to the Irish, as inherently un-British and racially threatening to both British appearance and character. As a Working Party report in 1953 concluded: 'Such a community is certainly no part of the concept of England or Britain to which people of British stock throughout the Commonwealth are attached.'[266] These beliefs mingled with perceptions of black sexual difference to create a deep well of objection to black residency in the UK. Other racial discourses cemented hostility towards the immigration of black communities, shaping attitudes towards black people as workers and neighbours.[267]

Partially driven by beliefs about the essential racial difference of new black immigrants, governments during the 1950s worked hard to limit the growing black population. By doing so they hoped to avoid the need for restrictive legislation, fearing that in the Commonwealth and the rest of the world, as well as at home, such an attempt would be perceived as racist. Thus a whole raft of surreptitious measures were employed to try and stem the flow. Early in the 1950s, the government attempted to alter African travel documents in order to obscure the nationality of the traveller on the form, thus removing his/her right to travel to Britain. New 'British travel certificates' were introduced in 1951 and 1952 ostensibly to prevent stowaways arriving in the UK. A Colonial Office memorandum recorded the ultimate intention of the policy.

> The West African colonies issue a special document, known as a British travel certificate, for local travel on the West African

coast, and these necessarily have to be issued freely without elaborate safeguards. Arrangements were however made to omit from these documents any statement as to nationality, so that any holder who arrived in the UK could be sent back under alien order powers.[268]

More substantially, attempts were made throughout the decade to influence Commonwealth countries into discouraging emigration to Britain and to spread propaganda in these countries about the lack of employment and housing opportunities in the UK. A 1958 minute from Colonial Office civil servant Ian Watt about his meeting with Parliamentary Under Secretary of State for the Colonies, John Profumo, recorded government intentions.

> After the meeting, Mr Profumo sent for me, to say that the suggestion had been aired in the meeting that there might be some value in HMG's trying to direct towards the West Indies some discreet propaganda to discourage unnecessary emigration to this country.[269]

Watt concluded by noting that a memorandum of this nature had been sent to every West Indian nation. It asked the governments in question to 'extend and expand their publicity measures to bring home to intending emigrants the present grave unemployment situation for West Indians in this country'.[270]

Attempts were also made to persuade other governments (notably India and Pakistan) to adopt a similar stance. Although these governments made a sincere effort to limit migration to Britain they were ultimately not able to prevent it, whilst Caribbean governments were much less cooperative about the idea of even trying to discourage their citizens from leaving.[271] The opportunity to migrate to Britain was important to their respective electorates and in nearly all cases domestic unemployment made governments reluctant to prevent the departure of excess workers. Additionally, there is evidence that some Commonwealth leaders resented British attempts to exclude their citizens, seeing restriction for the racist policy that it was. Speaking in Birmingham in June 1961, Norman Manley (the Prime Minister of Jamaica) argued the point:

> I have not been able to find any sound economic or social reason for putting a ban upon migrancy now and, therefore, I conclude

that the real reason – many people who live in England have come to the same conclusions – has to do with colour.²⁷²

Deakin has recorded the ultimate failure of this policy, concluding that 'the West Indian governments remained unprepared to take any responsibility for a decision which would certainly be exceedingly unpopular with their electorates'.²⁷³

Finally, having failed to exercise the desired level of control with administrative methods, the government moved to legislate. Fuelled by growing public unrest, which climaxed in racist rioting in Nottingham and Notting Hill Gate in 1958, work began on what would eventually become the Commonwealth Immigrants Act of 1962. It would be a landmark Act, which changed for good the concept of Britishness and the rights of the Commonwealth, created as the perceived least undesirable method of solving the black immigration problem.²⁷⁴

Most scholars of British black history have argued that the Commonwealth Immigrants Act was carefully and deliberately designed to exclude black immigrants. One analyst has concluded regarding the racial intention of the Act: 'It was how the country at large took it and was meant to take it.'²⁷⁵ That this question is even a matter of debate is due to the fact that the Conservative government, who created the legislation, emphatically denied that any form of racial restriction existed within it.²⁷⁶ The new law nominally applied to all Commonwealth countries and thus, it was argued, did not display any anti-black prejudice. However, overwhelming evidence leaves little doubt as to the Act's racial intentions.

The covert nature of the Act's racism reflects the British establishment's embarrassment about openly pursuing a racial policy.²⁷⁷ Essentially racist, the Act carefully avoided any overt racial language or intention. As Dummett and Dummett have concluded:

> ...to those least willing to acknowledge the presence of their own prejudices, it could be represented as not racial in character at all, but merely motivated by a concern for the total population figures in this 'crowded little island'. The fact that such a case was patently spurious did not matter at all.²⁷⁸

Among politicians and officials, the 'spurious' nature of the non-racialism in the Commonwealth Immigrants Act was well realised. However, the British politicians behind the legislation were very keen that it should be enacted beneath a cloak of non-racialism. As the law was finally drafted in 1961, a deluge of evidence confirms its hidden agenda. A memorandum for the Home Secretary highlights the point:

> We must recognise that although the scheme purports to relate solely to employment and to be non-discriminatory, its aim is primarily social and its operation is intended to, and would in fact, affect coloured people almost exclusively.[279]

The Inter-Departmental Working Party on Immigration reported the strengths of the coming bill to the Lord Chancellor:

> While it would apply equally in all parts of the Commonwealth, without distinction on grounds of race and colour, in practice it would interfere to the minimum extent with the entry of persons from the 'old' Commonwealth countries who would tend to come in categories (1) and (2).[280]

Put even more blatantly by the committee a few months later, the Bill would 'leave the door wide open' to white immigration.[281]

We can see in this legislation a political mirroring of the views and concerns of Britain's conservative biologists and eugenicists, who shared the government's desire to prevent as far as possible further black/white racial mixing in Britain. However, in the government's attempts to play ostensibly a non-racial hand we can perhaps also see the influence of Britain's social scientists, who were working to promote a social redefinition of race in this period.[282] But perhaps most of all, government and public views on black immigrants in this period reflected the thinking of Britain's progressive biologists, who had said so little on the subject of Commonwealth immigrants. Like the progressives, the government did not wish to associate with the forces of political racism and was keen to maintain non-racist credentials, yet its members were not inclined to ignore racial issues, or allow an open immigration policy which flouted the idea of racial difference by enabling the entry of an unlimited quantity of black people into Britain's racial stream. By the 1960s Britain was very slowly moving towards a post-racial era. However, complete abandonment of the idea of race remained a radical position. In what must with hindsight be seen as a transitional period, beliefs in racial difference continued to play a dominant role in political, social and scientific thinking.

5
Epilogue

Scientific beliefs in racial mental difference and racial hierarchies have been increasingly marginalised since the interwar period. Political battles against racial discrimination (especially Nazism) have kept the legitimacy of utilising race as a classificatory system in the dock of scientific as well as social consideration ever since this time. However, despite the best efforts of UNESCO and others, the idea of race never entirely disappeared from British biology in the post-war period. On the contrary, this study has argued that the concept survived in British science as both progressive and conservative biologists declined to dismiss race as a dividing mechanism. Politics led scientists to review racial analyses, to offer the public wide-ranging mitigations of race and to emphasise the errors of dogmatic science where it had been harnessed to causes of racial oppression. But politics did not destroy the biologists' belief in race.

After the war, the political implications of racial theories did make many British biologists, along with politicians and commentators, reluctant to be drawn on the subject. Far from trying to lead public opinion, as often they had done during the war, scientists became increasingly inward-looking and circumspect where they had previously been vocal. Only occasionally were leading scholars in the post-war period prepared to put their heads above the parapet on the issue of race. One man who saw fit to do so was the biologist J.R. Baker, whose book *Race* has been described by Kohn as 'possibly the last major statement of traditional racial science written in English'.[1] Baker's willingness to publish a major work defending the ideas of both racial hierarchy and racial mental difference was very unusual and seems to have stemmed from a personal determination to challenge the UNESCO position on racial equality, which he described in correspondence as 'a dogma not established by rigorous proofs'.[2] Although *Race* stood in isolation in its period, as a study it spoke something of what many within the British biological profession still thought but were afraid or unwilling to say.

Baker was a senior British academic figure, an Oxford University Reader in Cytology and long-standing eugenics enthusiast. However, when he decided to publish a book in defence of aspects of the idea of race he ran into a wall of opposition.[3] In correspondence with his friend and confidant C.P. Blacker, Baker admitted that his book proposal had been rejected by 'no fewer than eight' presses and that he had been 'subjected [by his literary agent] to the ultimate indignity of his refusal to even show the manuscript to a single publisher'.[4] This suspicion and hostility towards Baker's proposal were, as we have seen in this book, a reflection of a broader shift in social perceptions both about the concept of race and who should be considered an expert on the subject. At the end of the 1960s, most publishers did not want to get involved with a scholar of natural sciences who planned to explain race as a biological marker of inherent differences. Baker, like so many of his colleagues, appeared at what should have been the peak of his career to be on the brink of marginalisation.

To Baker's surprise his proposal was finally accepted by the prestigious Oxford University Press (OUP). He expressed his delight to Blacker: 'The Oxford University Press continues to astonish me. They are giving me a carte blanche contract. I'm allowed to write as much or as little as I like and what I like.'[5] The decision by OUP to offer such a generous contract to Baker seems to have been rooted in the support given to him by the Nobel prize-winning zoologist Sir Peter Medawar, who described his book in a personal reference to the publisher as 'a one-man Royal Commission Report on race'.[6] This praise, along with Baker's strong research background and other sympathetic reports, convinced OUP that Baker's opinions were worth publishing. After some wrangling over exact content, the book *Race* appeared in 1974.

Race offered a firm defence of the ideas of mental and physical racial difference.[7] Most fundamentally, it challenged the UNESCO doctrine of racial equality as scientifically unsound. Using the nomenclature of 'taxa', explained by the author as a categorisation of animal groups 'by their resemblances, and so far as possible by those resemblances that are due to common ancestry', Baker argued that it was unlikely in the context of evolutionary behaviour that human populations had developed mentally in the same way and at the same pace, given their different physiologies and environmental experiences:[8]

> Keeping in mind the close similarities of the skulls of jackal and fox, and the differences in their habits, one might find it difficult to suppose that two human taxa differing so profoundly in cranial characters as (for instance) the Eskimid and Lappid can be identical in all the genes that control their nervous and sensory systems; and if this seems unlikely, one must ask oneself whether it is conceivable that the mental qualities of each human taxon, though differing, must somehow add themselves together in such a way that all taxa are necessarily to be regarded as 'equal' mentally, in the special sense that no taxon is superior to any other. What known cause of evolution could have produced this result? Is it not more probable that natural selection has adapted taxa to different environments, and that as a result some of them have a greater tendency than others to produce persons possessing special agility and versatility of mind?[9]

In Baker's analysis, Australian aborigines served as an example of a 'taxa' which had intellectually developed to a lower degree than others, whilst the northern European, because of tougher environmental demands, had leapt ahead of other human groups.[10]

Baker went some way to distance his work from racist extremism. Like the progressives, he stressed the overlapping of racial traits and the role of environment in shaping human potential. He attacked the Nazi use of race and the practice of human slavery. Nonetheless, *Race* proffered a determined defence of conservative European racial study within biology and anthropology. Baker both utilised and lauded the work of theorists who had been widely dismissed by the post-war period as the forerunners of Hitler, such as the Comte de Gobineau and Houston Stewart Chamberlain. Baker indeed went as far as to assert:

> If one had to choose a single work as the most important of all in presenting one side of the ethnic controversy it would be reasonable to suggest Gobineau's *Essai sur l'inégalité des races humaines*; but necessarily it is in many respects out of date, and its very great length would deter most readers.[11]

Similar to the editors of the *Mankind Quarterly*, Baker presented his book as an impartial thesis on race, designed to challenge what he perceived as the politically motivated retreat from racial study which had occurred in the wake of the Third Reich. 'From the beginning of the thirties onwards scarcely anyone outside Germany and its allies dared to suggest that any race might be in any respect or in any sense superior to any other, lest it should appear that the author was supporting or excusing the Nazi cause.'[12] As we have seen in previous chapters, Baker was not alone in wishing that race had not fallen out of fashion in this way. To varying extents, pretty much all of Britain's leading biologists felt constrained by the UNESCO position, to which they mostly declined to subscribe. Baker's project was particularly welcomed by Britain's leading eugenicists. When he heard about the OUP contract, C.P. Blacker was so delighted that he felt 'like getting into my little red car and driving here and now to Kidlington with a couple of bottles of champagne and glasses and a corkscrew'.[13]

Like many other British biologists and eugenicists, Baker's pretensions of impartiality on the race issue were exaggerated. Whilst he no doubt did want to produce a comprehensive and scientific analysis of race, he retained strong political views on the subject which he was not keen to declare.[14] He responded to a leaflet that Blacker sent him from the Racial Preservation Society by noting his 'broad agreement'. Like Gates and his stance towards Southern segregationists, Baker declined to join formally the organisation only so that he could present himself as 'a perfectly free agent'.[15] Where it did not damage his academic credentials, Baker was prepared to stand alongside racial extremists in science. In 1973, he agreed to serve as a witness in support of the *Mankind Quarterly* editor Robert Gayre's action against the *Sunday Times*. Baker only agreed to appear in Court for Gayre after he was assured that the hearing would not precede the publication of *Race*, and thus would not reveal the tenor of his views before the book's arrival. After he was told by the publisher that the Court appearance would not occur prior to *Race*'s publication, Baker felt that he was honour bound to attend and defend a fellow race believer. He explained to Blacker:

> This being so, there was no longer any reason why I should not try and help Gayre, and I was troubled by the thought that Gayre might lose his case and I would go through the rest of my life with my conscience telling me that I had 'funked' (to use an old fashioned term) acting as a witness on his behalf.[16]

Although this correspondence suggests that Baker realised the political and academic ramifications of too close associations with political and scientific racists, it also reveals that he did not see his political stance as an important obstacle to his scientific impartiality. To Baker, and many of those scientists who waded into defences of race, there was no perceived contradiction between scientific objectivity and political ideology. Mostly, these scholars believed that their political views were shaped by scientific truth. Commenting on Gary Werskey's suggestion to him that political and social context shaped scientific opinion, Baker argued:

> Werskey thinks people's backgrounds have a profound influence on what they think. I tried to persuade him that some things are thought by some people because they happen to be true.[17]

It thus cannot be said with any certainty that Baker's scientific views on race were shaped by his politics. Instead it seems that he felt driven in science and politics alike simply by what he considered to be the objective truth. As this study has argued throughout, it is ultimately impossible to separate the two contexts in any case. Whatever his agenda, Baker's *Race* became the last major and mainstream British volume of the conservative racial biology which had been, in Barkan's parlance, 'retreating' since the publication of *We Europeans* 40 years before. That a book of its kind could be published after so long tells us much about the nature of the retreat of race in biology. It was slow, uncertain and divisive. However, the isolation of *Race*, and the difficulties that Baker had in bringing the project to fruition, tells us also that the times, by the 1970s, had changed. It has been argued here that social science, and not biology, led the way into a new post-racial scientific era and that biologists were mostly left unconvinced by bold claims of racial equality. Nonetheless, led by political will and a fear of the stigma increasingly attached to racism, British biology generally trimmed its sails to the new state of affairs so that Baker's *Race* stands out as an exception that proves the rule.

Whichever of science and politics took the lead in shaping the actions of scholars it has become clear that between 1930 and 1962 the study of race was riddled with political agendas. The conservative and progressive natural and social scientists all had their own political positions which undoubtedly had an impact on both their questions and answers about race. In British science and society after the Second World War, racism became increasingly taboo. In this atmosphere, scientific racial study attained something of a pariah status, perceived as scholarship which served to legitimise oppression and persecution. This did not mean that the racial ideas which had been part of the social and scientific ideological tapestry for generations disappeared in some kind of magic UNESCO wind. Instead, racial thinking survived, retained in new social discourses of cultural difference, in concerns about immigration and ethnic conflict. In biology, ideas of racial disparity (both mental and physical) also remained, smoothed over by apologetic language which emphasised the importance of the environment in shaping potential, the overlapping amongst different populations of all human traits and, perhaps most importantly, the idea of anti-racism.

Likewise, in our avowedly anti-racist twenty-first-century society, ideas of racial difference still lurk in the social and political construction of minorities and foreigners and in our understanding of ourselves. It is even arguable that Britain continues to operate an immigration policy which responds to notions of race within its scope. As was the case with refugees from Nazism, European Volunteer Workers and Commonwealth immigrants, immigration restrictions continue to be shaped by murky, racialised criteria of worth and 'belonging'.[18] Ultimately, whilst it has been reformed, reworked and refined, it seems that race will remain an influential idea for some time to come, entrenched in our social psyche. Because scientists function within this discursive terrain it seems probable that, in one guise or another, race will remain with them as long as it remains with us.

Notes

1 Introduction

1. Julian Huxley MSS, The Woodson Research Center, Rice University, Houston. Series 3, Box 18, Singer to Huxley, 22/4/49. The underlining here is Singer's.
2. As we shall see below, Singer and Huxley worked together to write antiracist books like *We Europeans* in 1935 and *Argument of Blood* in 1941.
3. See P. Gilroy, *Between Camps: Nations, Cultures and the Allure of Race* (London: Penguin, 2000), pp. 11–53 and C. Alexander, 'Beyond Black: Rethinking the Colour/Culture Divide', *Ethnic and Racial Studies* (2002), 25: 4, 552–71.
4. R. Miles, *Racism and Migrant Labour* (London: Routledge and Kegan Paul, 1982), pp. 32–3. Also see B. Carter, *Realism and Racism: Concepts of Race in Sociological Research* (London and New York: Routledge, 2000), pp. 164–6 and M. Banton, 'Progress in Ethnic and Racial Studies', *Ethnic and Racial Studies* (2001), 24: 2, 173–94.
5. Carter, *Realism and Racism*, p. 165.
6. Banton, 'Progress', p. 175.
7. As Miles put it: '..."race" is used in everyday discourse to refer to those aspects of physical variation which were used by nineteenth century science to identify permanent and discrete physical types', Miles, *Racism and Migrant Labour*, p. 20.
8. Gilroy, *Between Camps*, p. 53.
9. That this book was not actually written by Huxley and Haddon but by a team of writers including Charles Singer, Alexander Carr Saunders and Charles Seligman has been noted by Barkan and is discussed at length below. For this discussion, and an assessment of the thinking on race in *We Europeans* see Chapter 4.
10. K. Malik, *The Meaning of Race: Race, History and Culture in Western Society* (Basingstoke: Macmillan, 1996), p. 122.
11. N. Stepan, *The Idea of Race in Science: Great Britain 1800–1960* (London: Macmillan, 1992), p. 144. Also see H. Nowotny, P. Scott and M. Gibbons, *Re-Thinking Science: Knowledge and the Public in an Age of Uncertainty* (Cambridge: Polity, 2001) and H. and S. Rose, *Science and Society* (London: Allen Lane, 1969). These scholars argued that it was 'fallacious' to make 'any attempt to describe science as some sort of external agent acting upon a society and thus transforming it', p. 240.
12. Barkan and Miles amongst others have emphasised an existing yet uncertain passage of knowledge between science and society, described by Barkan as 'indirect and at times enigmatic'. See E. Barkan, *The Retreat of Scientific Racism: Changing Concepts of Race in Britain and the United States Between the World Wars* (Cambridge: Cambridge University Press, 1992), p. 10 and Miles, *Racism and Migrant Labour*, p. 19.

13 D. Goldberg, *Racist Culture: Philosophy and the Politics of Meaning* (Oxford: Blackwell, 1993), p. 45.
14 Barkan, *The Retreat of Scientific Racism*, p. 5.
15 M. Gibbons et al., *The New Production of Knowledge: the Dynamics of Science and Research in Contemporary Societies* (London: Sage, 1994), p. 22. Also see M. Hasian Jr., *The Rhetoric of Eugenics in Anglo-American Thought* (Athens: University of Georgia Press, 1996), p. 193, B. Carter, *Realism and Racism: Concepts of Race in Sociological Research* (London: Routledge, 2000), p. 138 and H. and S. Rose, *Science and Society*, p. 245.
16 Also see Gibbons, *The New Production of Knowledge*. As regards race, this tendency is evident in Michael Banton's analysis, where he argues that ideas of race are divisible into 'folk concepts' and 'scientific discourse'. See M. Banton, *Racial Theories* (Cambridge: Cambridge University Press, 1998) (first edition 1987), p. 3.
17 R. Smith, 'Biology and Values in Interwar Britain: CS Sherrington, Julian Huxley and the Vision of Progress', *Past and Present* (2003), 178, 210–42, 210–12.
18 W. McGucken, *Scientists, Society and the State: the Social Relations of Science Movement in Great Britain 1931–47* (Columbus: Ohio State University Press, 1984), p. 357.
19 Nowotny, Scott and Gibbons, *Re-Thinking Science*, p. 1. Also see Gibbons, *The New Production of Knowledge*.
20 Gibbons, *The New Production of Knowledge*, p. 4.
21 R. King, *Race, Culture and the Intellectuals, 1940–1970* (Baltimore: Johns Hopkins University Press, 2004), p. 1.
22 Barkan, *The Retreat of Scientific Racism*, pp. 1–3.
23 Stepan, *The Idea of Race*, pp. 141–3.
24 King, *Race*, p. 24.
25 See W. Tucker, *The Funding of Scientific Racism: Wickliffe Draper and the Pioneer Fund* (Urbana: University of Illinois Press, 2002) and M. Kohn, *The Race Gallery: the Return of Racial Science* (London: Jonathan Cape, 1995).
26 Gilroy, *Between Camps*, p. 25.
27 Malik, *The Meaning of Race*, p. 127.
28 Banton, *Racial Theories*, p. 6.
29 This change occurred, to Banton's mind, in the nineteenth century as the 'work of geologists forced the natural historians to reconsider their time-scale and to confront the evidence for change in all living forms', *Racial Theories*, p. 6.
30 Banton, *Racial Theories*, p. 7.
31 Banton, *Racial Theories*, p. 9. Also see Barkan, *The Retreat of Scientific Racism*, p. 140 and Stepan, *The Idea of Race*, pp. 120–1.
32 These kinds of fears as regards Jews were of course prominent in Nazi anti-Semitic propaganda. Concerns about the effects of black and white racial crossing were fairly prominent in European and American scientific racial discourse. Some of the more extreme examples of this kind of scientific analysis include C. Davenport and M. Steggarda, *Race Crossing in Jamaica* (Washington: Carnegie Institute, 1929), R. Gates, *Heredity and Eugenics* (London: Constable, 1923), J. Gregory, *The Menace of Colour*

(London: Seeley Service, 1925) and K. Aikman, 'Race Mixture', *Eugenics Review* (1933), 25:3, 161–6.

33 These theories were significant in shaping attitudes towards black people in the UK as well as in the United States. See P. Hoch, *White Hero Black Beast: Racism, Sexism and the Mask of Masculinity* (London: Pluto, 1979), pp. 43–64, Gilroy, *Between Camps*, pp. 196–7 and J. Hampshire, *Citizenship and Belonging: Immigration and the Politics of Demographic Governance in Postwar Britain* (Basingstoke: Palgrave Macmillan, 2005), pp. 111–49.

34 For most of the scientists under discussion, research on race was not their primary scientific concern, a point which Barkan has made in *The Retreat of Scientific Racism*, p. 5. Perhaps the only exception to this rule was Reginald Gates, who became so obsessed by race that racial enquiries made up the bulk of his work from the 1940s onwards.

35 G. Werskey, *The Visible College: a Collective Biography of British Scientists and Socialists of the 1930s* (London: Allen Lane, 1978).

36 For recent scholarship on the British Eugenics Society see P. Mazumdar, *Eugenics, Human Genetics and Human Failings: the Eugenics Society, its Sources and its Critics in Britain* (London and New York: Routledge, 1992), R. Soloway, *Demography and Degeneration: Eugenics and the Declining Birthrate in Twentieth Century Britain* (Chapel Hill and London: The University Press of North Carolina, 1990), R. Peel (ed.), *Essays in the History of Eugenics* (London: The Galton Institute, 1997), D. Kevles, *In the Name of Eugenics: Genetics and the Use of Human Heredity* (New York: Knopf, 1985) and G. Jones, *Science, Politics and the Cold War* (London and New York: Routledge, 1988), pp. 61–2.

37 D. Stone, 'Race in British Eugenics', *European History Quarterly* (2001), 31: 3, 397–425. For the idea that racial ideas were 'incidental' see D. MacKenzie, 'Eugenics in Britain', *Social Studies of Science* (1976), 6: 3–4, 499–532, 501.

38 Stack has recently highlighted the important role played by the left in the development of British eugenics in its early period in D. Stack, *The First Darwinian Left: Socialism and Darwinism 1859–1914* (Cheltenham: New Clarion Press, 2003). Kevles has noted the ongoing significance of liberals and leftists in the 1930s reform agenda of C.P. Blacker in *In the Name of Eugenics*, pp. 164–75. Also see Mazumdar, *Eugenics*, p. 185.

39 R. Soloway, 'From Mainline to Reform Eugenics – Leonard Darwin and CP Blacker', in Peel, *History of Eugenics*, pp. 52–80.

40 Hasian has noted in this context that the term 'eugenics' was utilised in eight entirely different ways in Hasian, *The Rhetoric of Eugenics*, pp. 28–9.

41 N. Stepan, *The Idea of Race in Science*, p. 111. For Galton's views on race see F. Galton, 'Hereditary Talent and Character', *Macmillan's Magazine* (1865), 12, 318–27.

42 C. Saleeby, *The Eugenic Prospect: National and Racial* (London: Fisher Unwin, 1921), Attitude to alcohol, p. 39, National Health Centres, p. 237.

43 A. White, *Efficiency and Empire* (London: Methuen, 1901), pp. 111–17.

44 Jones, *Science, Politics and the Cold War*, p. 62.

45 Smith, 'Biology and Values in Interwar Britain', p. 238.

46 See B. Gainer, *Alien Invasion: the Origins of the Aliens Act of 1905* (London: Heinemann, 1972), B. Porter, *The Refugee Question in Mid Victorian Politics*

(Cambridge: Cambridge University Press, 1979), p. 218, C. Holmes, *John Bull's Island: Immigration and British Society 1871–1971* (Basingstoke: Macmillan, 1988), pp. 71–4 and V. Bevan, *The Development of British Immigration Law* (London: Croom Helm, 1986), pp. 67–72.

47 *Hansard*, Vol: 145, Col: 724, 2/5/05. For analysis of the provisions of the Act and its scope see Gainer, *Alien Invasion*, pp. 191–2.

48 For origins of the society see Mazumdar, *Eugenics*, pp. 7–57 and G. Searle, 'Eugenics: the Early Years' in Peel, *Essays*, pp. 20–35. For the agenda behind the Aliens Act see C. Holmes, *Anti-Semitism in British Society 1876–1939* (London: Edward Arnold, 1979), p. 37 and Gainer, *Alien Invasion*, pp. 166–98.

49 For contemporary accounts of this sort, see A. White, *Efficiency and Empire*. For analysis see G. Searle, who explores the views of the eugenicist Karl Pearson on degeneration in *The Quest for National Efficiency: a Study in British Politics and Political Thought 1899–1914* (Oxford: Blackwell, 1971), p. 39 and M. Rosenthal, *The Character Factory: Baden Powell and the Origins of the Boy Scout Movement* (New York: Pantheon, 1986), pp. 131–60.

50 See M. Hawkins, *Social Darwinism in European and American Thought: Nature as Model and Nature as Threat 1860–1945* (Cambridge: Cambridge University Press, 1997), pp. 184–215.

51 D. Cesarani, 'An Alien Concept? The Continuity of Anti-Alienism in British Society before 1940' in D. Cesarani and T. Kushner, *The Internment of Aliens in Twentieth Century Britain* (London: Frank Cass, 1993), pp. 25–52.

52 *Hansard*, Vol: 145, Col: 740, 2/5/05.

53 *Hansard*, Vol: 145, Col: 796, 2/5/05. For analysis see Gainer, *Alien Invasion*, pp. 166–98.

54 The new law was the Aliens Restriction Act. See T. Kushner and K. Knox, *Refugees in an Age of Genocide: Global, National and Local Perspectives during the Twentieth Century* (London: Frank Cass, 1999), p. 44 and C. Holmes, *John Bull's Island*, pp. 94–5.

55 See P. Panayi, *Enemy in our Midst: Germans in Britain during the First World War* (Oxford: Berg, 1991).

56 Aliens Restriction (Amendment) Act 1919, Articles 10 and 11.

57 See Kushner and Knox, *Refugees in an Age of Genocide*, pp. 43–99.

58 *Hansard*, Vol: 114, Cols: 2785 and 2799, 15/4/19.

59 1923 Speech by Charles Crook, Conservative MP for East Ham. Cited in P. Foot, *Immigration and Race in British Politics* (London: Penguin, 1965), p. 110. These comments match almost to the word the ideas of Reginald Ruggles Gates in 1923. See Gates, *Heredity and Eugenics*, pp. 237–9.

60 Aliens Restriction Amendment Act, 1919. Article 7 prohibited alien name changes, Article 11 prohibited any alien ownership of land or business for three years, whilst Article 3 dealt with punishments for aliens attempting to cause 'sedition or disaffection'.

61 See B. Cheyette, *Constructions of 'the Jew' in English Literature and Society: Racial Representations 1875–1945* (Cambridge: Cambridge University Press, 1993), M. Ragussis, *Figures of Conversion: 'the Jewish Question' and English National Identity* (Durham and London: Duke University Press, 1995) and D. Feldman (ed.), *Englishmen and Jews: Social Relations and Political Culture 1840–1914* (New Haven and London: Yale University Press, 1994).

62 See S. Kadish, *Bolsheviks and British Jews: the Anglo-Jewish Community and the Russian Revolution* (London: Frank Cass, 1992), pp. 10–73.
63 Aliens Restriction Amendment Act, 1919, Article 5/2.
64 For analysis of these Orders see L. Tabili, 'The Construction of Racial Difference in Twentieth Century Britain: the Special Restriction (Coloured Alien Seamen) Order, 1925', *Journal of British Studies* (1994), 33, 54–98, P. Rich, *Race and Empire in British Politics* (Cambridge: Cambridge University Press, 1986), pp. 122–30 and C. Holmes, *A Tolerant Country? Immigrants, Refugees and Minorities in Britain* (London: Faber & Faber, 1991), p. 36.
65 Aliens Order, 25/3/20.
66 Special Restriction Order, 18/3/25.
67 See the retrospective analysis of the impact of the legislation on Black seamen, in *The Keys: the Journal of the League of Coloured Peoples* (1935), 3: 1, 4–22.
68 See P. Fryer, *Staying Power: the History of Black People in Britain* (London: Pluto, 1984), pp. 298–312 and N. Evans, 'Regulating the Reserve Army: Arabs, Blacks and the Local State in Cardiff, 1919–45' in K. Lunn, *Race and Labour in Twentieth Century Britain* (London: Frank Cass, 1985), pp. 68–115.
69 *The Keys* (1935), 3: 1, 4.
70 *The Keys* (1935), 3: 2, 19.
71 PRO, ADM 1/8696/40, Meeting of Aliens and Nationality Committee, 5/8/21.
72 PRO, HO 73/112, Department of Overseas Trade Memorandum, 20/5/19.
73 *Jewish Chronicle*, 11/7/19.

2 Rethinking Interwar Racial Reform: the 1930s

1 See Mazumdar, *Eugenics*, pp. 146–95, Stepan, *The Idea of Race*, pp. 140–69, Barkan, *The Retreat of Scientific Racism*, pp. 279–340 and Rich, *Race and Empire*.
2 Stepan, *The Idea of Race*, p. 141, Mazumdar, *Human Genetics*, p. 146.
3 King, *Race, Culture and the Intellectuals*, p. 1.
4 The use of the term 'race' did become controversial during the decade as several scholars suggested that the term should be dropped altogether. However, this chapter will argue that this call for change mainly represented a politically driven semantic disaffection with the word 'race', not any substantial call for the dismissal of the idea.
5 See M. Moul and K. Pearson, 'The Problem of Alien Immigration into Great Britain Illustrated by an Examination of Russian and Polish Alien Children', *Annals of Eugenics* (1925–6), 1, 6–127. Pearson was Professor in Mathematics at University College London until his retirement in 1933. He himself founded the *Annals of Eugenics* in 1925.
6 M. Davies and A.G. Hughes, *An Investigation into the Comparative Intelligence and Attainments of Jewish and Non-Jewish School Children* (Cambridge: Cambridge University Press, 1927). For analysis of the funding of this research see G. Richards, *'Race', Racism and Psychology: Towards a Reflexive History* (London and New York: Routledge, 1997), p. 191. A.G. Hughes elaborated on this research in A.G. Hughes, 'Jews and Gentiles: their Intellectual and Temperamental Differences', *Eugenics Review* (1928), 20: 2, 89–94. Also

see T. Endelman, 'Anglo-Jewish Scientists and the Science of Race', *Jewish Social Studies* (2004), 11: 1, 52–92, 78–9.

7. M. Fletcher, 'Report on an Investigation into the Colour Problem in Liverpool and Other Ports', Liverpool Central Archive (File H325 26 FLE).
8. See P. Rich, 'Philanthropic Racism in Britain: the Liverpool University Settlement, the Anti-Slavery Society and the Issue of "Half-Caste" Children, 1919–51', *Immigrants and Minorities* (1984), 3: 1, 69–88, 71, C. Wilson, 'Racism and Private Assistance: the Support of West Indian and African Missions in Liverpool, England, During the Interwar Years', *African Studies Review* (1992), 35: 2, 55–76, 62–3 and Rich, *Race and Empire*, pp. 132–6. Also see C. King and H. King, *'The Two Nations': the Life and Work of the Liverpool University Settlement and its Associated Institutions 1906–37* (London: Liverpool University Press, 1938), pp. 127–34.
9. See D. Stone, 'Race in British Eugenics', *European History Quarterly* (2001), 31: 3, 397–425, 415.
10. The Committee was chaired by Prof. Roxby of the School of Geography at the University of Liverpool. See Rich, 'Philanthropic Racism', p. 72.
11. D. Caradog Jones, *The Economic Status of Coloured Families in the Port of Liverpool* (Birkenhead: Woolman and Sons, 1940). For analysis see Rich, 'Philanthropic Racism', p. 74.
12. Pearson wrote up the beginnings of this report for his personal notes in 1911 as a 'Report on the Investigations in Progress at the Jews Free School, July to December 1910'. K. Pearson MSS, University College London Archive, MSS 140–1, 1/1/11.
13. Pearson MSS, 140–1, 1/1/11.
14. Moul and Pearson, 'Alien Immigration', p. 7.
15. For recent analysis of this research see Barkan, *The Retreat of Scientific Racism*, pp. 155–7, Richards, *'Race', Racism and Psychology*, pp. 191–3 and G. Schaffer, '"Like a Baby with a Box of Matches": British Scientists and the Concept of "Race" in the Interwar Period', *British Journal for the History of Science* (2005), 38: 3, 307–24, 315–17.
16. Moul and Pearson, 'Alien Immigration', p. 9. The authors wrote of Lord Rothschild: 'Both he and other English Jews admitted the gravity of our problem by contributing to the fund we had to raise in order to carry through the work.'
17. For analysis of immigration debates in this period see Holmes, *John Bull's Island*, Bevan, *The Development of British Immigration Law*, Cesarani, 'An Alien Concept?, D. Cesarani, 'Anti-Alienism in England after the First World War', *Immigrants and Minorities* (1987), 6: 1, 5–29 and Kushner and Knox, *Refugees in an Age of Genocide*, pp. 64–102.
18. Moul and Pearson, 'Alien Immigration', p. 51.
19. Moul and Pearson, 'Alien Immigration', p. 126.
20. Moul and Pearson, 'Alien Immigration', p. 127.
21. Moul and Pearson, 'Alien Immigration', p. 124.
22. Richards has argued that few of Moul and Pearson's contemporaries agreed with their analysis. See Richards, *'Race', Racism and Psychology*, pp. 192–3.
23. Davies and Hughes, *An Investigation*, p. 146. For analysis of the funding of this research see Endelman, 'Anglo-Jewish Scientists', p. 79.
24. Davies and Hughes, *An Investigation*, p. 140.

25 Davies and Hughes, *An Investigation*, p. 145.
26 Richards, *'Race', Racism and Psychology*, p. 191.
27 Davies and Hughes, *An Investigation*, p. 145. This emphasis on environment as the primary factor in immigrant success was even more strongly expressed in Rumyaneck's criticism of research into Jewish intelligence. See J. Rumyaneck, 'The Comparative Psychology of Jews and non-Jews: a survey of the literature', *The British Journal of Psychology* (1931), 11: 4, 404–23. Rumyaneck's analysis of this research was commissioned, like the Davies and Hughes report, by the *Jewish Health Organisation of Great Britain*. For details see Endelman, 'Anglo-Jewish Scientists', p. 79.
28 The children in their survey were measured using the Northumberland Standardized Tests, prepared by Burt himself. Davies and Hughes, *An Investigation*, p. 135. In the final line of the report the authors thanked Burt for his 'helpful advice and guidance'.
29 For analysis of Burt's career see L. Hearnshaw, *Cyril Burt Psychologist* (London: Hodder and Stoughton, 1979) and S.J. Gould, *The Mismeasure of Man* (London: Penguin, 1981), pp. 234–55.
30 Stepan, *The Idea of Race*, p. 133.
31 Hughes, 'Jews and Gentiles', p. 89.
32 Hughes, 'Jews and Gentiles', p. 94.
33 Hughes, 'Jews and Gentiles', p. 94.
34 For analysis see J. Jenkinson, 'The Glasgow Race Disturbances of 1919', in Lunn (ed.), *Race and Labour*, pp. 43–67, J. Jenkinson, 'The Black Community of Salford and Hull 1919–21', *Immigrants and Minorities* (1988), 7: 2, 166–83, N. Evans, 'The South Wales Race Riots of 1919', *Llafer* (1983), 3, 5–29 and Fryer, *Staying Power*, pp. 298–307.
35 See A. Murphy, *From the Empire to the Rialto: Racism and Reaction in Liverpool 1918–48* (Birkenhead: Liver Press, 1995), pp. 17–18 and M. Sherwood, 'Lynching in Britain', *History Today* (1999), 49: 3, 21–3.
36 See Murphy, *From the Empire*, pp. 28–42.
37 See Rich, 'Philanthropic Racism', pp. 75–88 and Wilson, 'Racism and Private Assistance', pp. 58–63.
38 See C. King and H. King, *'The Two Nations': the Life and Work of the Liverpool University Settlement and its Associated Institutions 1906–37* (London: Liverpool University Press, 1938).
39 King and King, *Two Nations*, p. 129.
40 King and King, *Two Nations*, p. 129. For analysis of Fleming's earlier research see Barkan, *The Retreat of Scientific Racism*, pp. 58–9.
41 'Fletcher Report', pp. 1–6.
42 'Fletcher Report', p. 8.
43 'Fletcher Report', p. 38.
44 'Fletcher Report', pp. 19–20.
45 'Fletcher Report', p. 23.
46 'Fletcher Report', p. 21.
47 'Fletcher Report', p. 23.
48 'There is little harmony between the parents, the coloured men in general despise the women with whom they consort, while the majority of women have little affection for the men...all the circumstances of their lives tend to give undue prominence to sex', 'Fletcher Report', p. 26.

49 'Fletcher Report', p. 15.
50 'Fletcher Report', pp. 33–5. Unsurprisingly this report generated bitterness and hostility. According to one contemporary analyst, Fletcher was threatened with physical violence in Liverpool in the wake of the report's publication. See King and King, *Two Nations*, p. 130.
51 Caradog Jones, *The Economic Status*, p. 5.
52 Caradog Jones, *The Economic Status*, p. 5.
53 Caradog Jones, *The Economic Status*, p. 8.
54 Caradog Jones, *The Economic Status*, p. 8.
55 Caradog Jones, *The Economic Status*, p. 7
56 See Rich, 'Philanthropic Racism', pp. 83–4.
57 Caradog Jones, *The Economic Status*, pp. 19–23.
58 Caradog Jones, *The Economic Status*, pp. 12–23.
59 Rich, 'Philanthropic Racism', p. 72.
60 See Werskey, *The Visible College*. Also Barkan, *The Retreat of Scientific Racism*, p. 142.
61 Mazumdar, *Eugenics*, p. 3.
62 See Mazumdar, *Eugenics, Human Genetics*..., G. Searle, *Eugenics and Politics in Britain 1900–14* (Leyden: Noordoff International Publishing, 1976) and G. Jones, 'Eugenics and Social Policy Between the Wars', *The Historical Journal* (1982), 25: 3, 717–28. Also see Richards, *'Race', Racism and Psychology*, pp. 189–216.
63 See Stepan, *The Idea of Race*, p. 140.
64 J.B.S. Haldane, *Science and Everyday Life* (Harmondsworth: Penguin, 1939), p. 178.
65 L. Hogben, *Dangerous Thoughts* (London: Allen and Unwin, 1939), p. 51.
66 Stone, 'Race in British Eugenics', p. 398. Also see D. Stone, *Breeding Superman: Nietzsche, Race and Eugenics in Edwardian and Interwar Britain* (Liverpool: Liverpool University Press, 2002).
67 Barkan, *The Retreat of Scientific Racism*, p. 1.
68 M. Kohn, *A Reason for Everything: Natural Selection and the English Imagination* (London: Faber and Faber, 2005), p. 162.
69 Werskey, *Visible College*. p. 241. Barkan has written three separate sections for Huxley in *The Retreat of Scientific Racism*: one recording his earlier racist thinking, the second his perspective in *Africa View*, the last, his famous change of heart in the face of Nazism in *We Europeans*. Barkan, *The Retreat of Scientific Racism*. See sections: 'A Racist Liberal. Julian Huxley's Early Years', pp. 177–89, '"Africa View" – Huxley's Changing Perspectives', pp. 235–49, and 'We Europeans', pp. 296–302.
70 Stepan, *The Idea of Race*, p. 143 and Barkan, *The Retreat of Scientific Racism*, p. 279.
71 'Head measuring' was perhaps the most prominent feature of this kind of racial science after the 'cephalic index' was invented by Swedish anatomist Anders Retzius in 1844. In this method people were categorised according to skull shape, long headed (dolichocephalic) and wide headed (brachycephalic). Populations were analysed in this way, notably in J. Beddoe, *The Races of Britain: a Contribution to the Anthropology of Western Europe* (London: Trubner and Co., 1885). Beddoe also conducted racial analysis according to eye colour, hair colour and body size and shape. Also see R. Knox, *The Races of Men* (London: Savill and Edwards, 1850).

72 See Stepan, *The Idea of Race*, pp. 134–8.
73 K. Pearson, *The Groundwork of Eugenics* (London: University College London Press, 1909), p. 10.
74 Indeed Stepan has argued that this critique had little initial impact. See Stepan, *The Idea of Race*, p. 134.
75 Pearson, *The Groundwork of Eugenics*, p. 10.
76 See Pearson's contribution to the Anthropometric Standards Committee cited in Rich, *Race and Empire*, p. 109.
77 Barkan, *The Retreat of Scientific Racism*, pp. 139–40.
78 L. Hogben, 'The Origins of the Society', in M.A. Sleigh and J.F. Sutcliffe (eds), *The Origins and History of the Society for Experimental Biology* (London: The Society for Experimental Biology, 1966), p. 5.
79 Hogben, 'The Origins of the Society', p. 7. Interestingly, Reginald Gates (who by the end of the 1930s was barely on speaking terms with Huxley because of their differing racial views) was also a board member, p. 8.
80 Lancelot Hogben, *Genetic Principles in Medicine and Social Science* (London: Williams and Norgate, 1931), pp. 122–44, p. 122.
81 Hogben, *Genetic Principles*, p. 133. In the seminal anti-Nazi text *We Europeans*, discussed later in this chapter, Mendelian reasoning was utilised at the core of the anti-racial argument. Analysing race in modern European populations the authors noted: 'after a cross the resulting population will not tend to a mere average between the two original ingredients, but will, in the absence of social or natural selection, continue to produce a great diversity of types, generation after generation'. J. Huxley and A. Haddon, *We Europeans: a Survey of Racial Problems* (London: Jonathan Cape, 1935).
82 Stepan, *The Idea of Race*, pp. 120–1.
83 Barkan, *The Retreat of Scientific Racism*, p. 140.
84 See *Race and Culture* (London: Royal Anthropological Institute and the Institute of Sociology, 1935), p. 2. Barkan has written in detail on this conference in Barkan, *The Retreat of Scientific Racism*, pp. 285–96.
85 See Endelman, 'Anglo-Jewish Scientists', p. 83.
86 *Race and Culture*, pp. 5–6. These views were consistent with Elliot Smith's published analysis on the subject. See G. Elliot Smith, *Human History* (London: Cape, 1930) and *The Diffusion of Culture* (London: Watts and Co, 1933).
87 *Race and Culture*, p. 19.
88 *Race and Culture*, p. 8. Haldane argued: 'some characters are little influenced by environment, but are innate'.
89 *Race and Culture*, p. 10.
90 Nazi thinking on racial type should not be oversimplified. For a sophisticated analysis see C. Hutton, *Race and the Third Reich: Linguistics, Racial Anthropology and Genetics in the Dialectic of Volk* (Cambridge: Polity, 2005).
91 See Barkan, *The Retreat of Scientific Racism*, p. 293.
92 Commenting on Barzun's 1938 book, 'Race: a Study in Modern Superstition', Gates noted: 'Jewish troubles in Germany etc really result from their refusal to be absorbed.' Gates MSS, File 11/41. For analysis of the racial views of Gates see G. Schaffer, '"Scientific" Racism Again?: Reginald Gates, the *Mankind Quarterly* and the Question of "Race" in Science after the Second World War', *Journal of American Studies* (2007), 41: 2, 253–78.

93 Pitt-Rivers was considered by some leading eugenicists in this period to be something of a loose cannon. In correspondence between the secretary of the Eugenics Society C.P. Blacker and Julian Huxley, Blacker voiced his reluctance to allow Pitt-Rivers to debate with G.K. Chesterton, noting: 'I think it very likely that GK Chesterton would wipe the floor with him – make a fool of him and cause him to lose his temper. When he does this, he bellows ghastly.' Julian Huxley MSS, Series 3, Box 11, Blacker to Huxley, 3/2/32.
94 Barkan has noted that leading Jewish scientists were excluded from the conference as it was argued that they could not consider issues of race without bias. See Barkan, *The Retreat of Scientific Racism*, p. 289.
95 Barkan, *The Retreat of Scientific Racism*, p. 295.
96 *Race and Culture*. Challenge to Elliot Smith, p. 18. On racial types, pp. 16–17.
97 This argument was intellectually rooted in polygenist racial theory, the idea that human beings had emerged from more than one original pair. The theory was most popular from the late eighteenth to the mid nineteenth century but retained some key supporters into the twentieth century. Notably, Arthur Keith's eager analysis of the later discredited 'Piltdown Man' famously carried a form of this thinking into the interwar period. See J. Sawday, '"New Men, Strange Faces, Other Minds": Arthur Keith, Race and the Piltdown Affair (1912–52)', in W. Ernst and B. Harris (eds), *Race, Science and Medicine, 1700–1960* (London: Routledge, 1999), pp. 259–88. Gates posited polygenist theory until his death in 1962. Indeed, these beliefs were recorded in his final academic contribution. See R. Gates, *The Emergence of Racial Genetics* (New York: International Association for the Advancement of Ethnology and Eugenics, 1960). For analysis see Schaffer, 'Scientific Racism Again'.
98 *Race and Culture*, p. 13.
99 *Race and Culture*, p. 15. As we shall see below, the argument that nations were evolving races was central to the racial world view of Gates's closest ally Arthur Keith.
100 *Race and Culture*, p. 17.
101 J. Huxley, *Memories I* (London: Allen and Unwin, 1970), p. 207.
102 See Barkan, *The Retreat of Scientific Racism*, pp. 296–310, Schaffer, 'Like a Baby...', pp. 309 and 321–2.
103 Julian Huxley MSS, 3/11, Coomb to Huxley, 8/1/35. The Pinker referred to in this correspondence was Huxley's agent, James B. Pinker.
104 Huxley MSS, 3/11, Coomb to Huxley, 22/1/35.
105 Huxley MSS, 3/11, Coomb to Huxley, 29/1/35.
106 The extent of collaboration in *We Europeans* is clear from the record of paid royalties. In a letter from the accountant E.S.P. Haynes to Huxley, Haynes confirms (concerning Huxley's tax return): 'The amounts paid to your collaborators were duly allowed in the assessment.' Huxley MSS, 11/15, Haynes to Huxley, 17/3/41.
107 Barkan, *The Retreat of Scientific Racism*, p. 306.
108 Gates MSS, Haddon to Gates, 23/2/37.
109 Huxley and Haddon, *We Europeans*, p. 92.
110 J. Huxley, *Africa View* (London: Chatto and Windus, 1931), p. 389.

111 Singer expressed this view in correspondence with biologist, Redcliffe Salaman. Singer criticised Salaman for presenting political Zionism as 'an impulse of persons with a particular type of nose...to huddle together'. Charles Singer MSS, Wellcome Archive, Wellcome Institute, London (PPCJS/E1), Singer to Salaman, 2/11/43. Singer's correspondence with doctor and eugenicist Frederick Parkes Weber further illuminates Singer's views on the need for the deconstruction of racial thinking. Singer wrote to Parkes Weber in 1943, criticising the doctor and citing the intellectual bankruptcy of racial Jewish analysis in a tone suspiciously similar to that adopted in *We Europeans*: 'Passing over the term "race", which seems to me to be biologically untenable in the racial group known as Jews and passing over the point that you treat the "Jewish nose" as a Mendelian dominant for which there is no evidence and against which there is much evidence – you make the oddly contradictory point that as Jewish "blood" gets more and more lost in the general population so jealousy of Jews grows more and more.' Parkes Weber MSS, Wellcome Institute, London, PP/FPW/C10, 6/10/43.
112 C. Dover, 'We Europeans', *Nature* (1935), 136: 3445, 736–7.
113 Dover, 'We Europeans', p. 736.
114 Huxley and Haddon, *We Europeans*, p. 18.
115 Huxley and Haddon, *We Europeans*, p. 27.
116 Huxley and Haddon, *We Europeans*, p. 273.
117 Huxley and Haddon, *We Europeans*, p. 136 and p. 268.
118 Huxley and Haddon, *We Europeans*, p. 283.
119 Huxley and Haddon, *We Europeans*, p. 287.
120 Huxley MSS, Series 3, Box 12, Gates to Huxley, 12/3/37.
121 See Barkan, *The Retreat of Scientific Racism*, pp. 303–5.
122 Huxley MSS, Series 3, Box 12, Haddon to Huxley, 9/8/35 and 15/9/35.
123 Huxley MSS, Series 3, Box 12, Gates to Huxley, 24/3/37. Barkan cites the letter from Haddon to Gates in full in *The Retreat of Scientific Racism*, p. 303.
124 Julian Huxley MSS, Series 3, Box 12, Gates to Huxley, 22/2/37.
125 Haddon had written extensively on race without challenging substantially the validity of the idea. See especially A. Haddon, *The Races of Man* (Cambridge: Cambridge University Press, 1924). His papers are retained in the Central University Library in Cambridge.
126 Huxley and Haddon, *We Europeans*, p. 281.
127 Huxley and Haddon, *We Europeans*, p. 91.
128 Huxley and Haddon, *We Europeans*, p. 109.
129 Huxley and Haddon, *We Europeans*, p. 283.
130 Huxley and Haddon, *We Europeans*, pp. 282–3.
131 Dover, 'We Europeans', p. 737.
132 For an analysis of this confusion see E. Barkan, 'The Dynamics of Huxley's Views on Race and Eugenics' in C.K. Waters and A. Van Helden, *Julian Huxley: Biologist and Statesman of Science* (Houston: Rice University Press, 1992), pp. 230–7.
133 See Kevles, *In the Name of Eugenics*, pp. 164–75, Soloway, *Demography and Degeneration*, pp. 193–225 and Soloway, 'From Mainline to Reform Eugenics', pp. 52–80.
134 Soloway, 'From Mainline to Reform Eugenics', p. 65 and Kevles, *In the Name*, p. 173.

135 As early as 1933, Huxley wrote to Blacker on the importance of British eugenic opposition to Nazism. See Eugenics Society MSS, SA EUG C185, Huxley to Blacker, 29/5/33. Kevles has noted that whilst Nazism was influential in driving the reform of the Eugenics Society, it paralleled an ongoing longer-standing process of introspection inside British eugenics. See Kevles, *In the Name*, p. 118.

136 After receiving a £70,000 bequest from an Australian sheep farmer named Henry Twitchin the Society funded the Birth Control Investigation Committee (which it had established in 1928), the British Social Hygiene Council, the Marriage Guidance Council, the National Birth Control Association (and its successor the Family Planning Association) and the Joint Committee on Voluntary Sterilisation according to Soloway, 'From Mainline to Reform Eugenics', pp. 64–8.

137 See Stepan, *The Idea of Race*. On Huxley, Haldane and Hogben in the 'anti-eugenic camp', p. 146. On Huxley and Haldane's continuing involvement, pp. 154–5. For further analysis of Haldane's attitude towards eugenics see Werskey, *The Visible College*, pp. 96–7.

138 Huxley gave the Galton lecture at the Eugenics Society in London on 6/6/1962. This lecture was printed in the *Eugenics Review*, 1962, 54: 123 and formed the basis of an article in *Nature*. See J. Huxley, 'Eugenics in Evolutionary Perspective', *Nature* (1962), 195: 4838, 227–8.

139 Saul Dubow has emphasised Hogben's ambivalent attitudes towards race in a discussion of his work in South Africa. See S. Dubow, *Scientific Racism in Modern South Africa* (Cambridge and New York: Cambridge University Press, 1995), pp. 191–4.

140 Eugenics Society MSS, SA EUG C163. Hogben served on the Birth Control Investigation Committee, before resigning in 1934 due to ill health, Hogben to Blacker, 7/11/34. In 1936 he agreed to serve on the Population Investigation Committee, Hogben to Blacker, 12/2/36.

141 Lancelot Hogben MSS, A84, 'The fifth year in the work of the department of social biology', 1935.

142 C.P. Blacker MSS, Box 25, Galton Lecture, 16/2/45. It is significant that G.P. Wells, in his biography of Hogben, has specifically highlighted Hogben's desire to attack 'the political and social bias of current eugenics', not the discipline itself. See G.P. Wells, *Lancelot Thomas Hogben: Biographical Memoirs of Fellows of the Royal Society* (London: The Royal Society, 1978), p. 197. Also see L. Hogben, *Dangerous Thoughts*. Hogben argued: 'The eugenic movement of this country has always been, and still remains, an organisation of a small section of the professional class with a strongly conservative bias directed to restrict the further extension of educational opportunities', p. 57.

143 Blacker MSS, Box 12/C1/1. Crew and Haldane received research funding from the BCIC. File 12/C3/2 shows membership lists including Crew, Carr Saunders, Huxley and Hogben in 1934. The Eugenics Society played a central role in political and social discussion of demographic issues in this period, notably setting up and funding the Population Investigation Committee in 1936. For details see E. Grebenik, 'Demographic Research in Britain 1936–86', *Population Studies* (1991), 45, pp. 3–30, p. 9.

144 Huxley MSS, Box 100, 'Aryan Myth Exploded', *San Francisco Examiner*, October 1938.

145   Huxley, 'Aryan Myth Exploded'.
146   J. Huxley, 'Galton Lecture', *Eugenics Review* (1936), 28: 1, 11–31, 17–18.
147   Huxley, 'Galton Lecture', p. 28.
148   Huxley, 'Galton Lecture', p. 13.
149   Huxley, 'Galton Lecture', p. 14.
150   Huxley calls for cooperation with sociology, 'Galton Lecture', p. 31.
151   Huxley MSS, Box 12, Drummond Shields to Huxley, 5/6/36.
152   Huxley MSS, Box 12, Goodrich to Huxley, 3/6/36. H.J. Muller was an American geneticist who in this period supported the social application of eugenic principles.
153   J.B.S. Haldane MSS, MS20604, 'Essay on Race' (1945/1946).
154   J.B.S. Haldane, *Heredity and Politics* (London: Allen and Unwin, 1938), p. 129. Haldane also retained a belief in important physical racial differences. In a 1925 study he argued that black soldiers should be utilised in future European conflicts on the grounds that they were largely immune to mustard gas attacks: '...it should be possible to obtain coloured troops who would all be resistant to mustard gas blistering in concentrations harmful to most white men. Enough resistant whites are available to officer them', J.B.S. Haldane, *Callinicus: a Defence of Chemical Warfare* (London: Kegan Paul, 1925), pp. 45–6. Not only does this idea reveal that Haldane believed that one human population could be insusceptible to gas attack where another was not, but it also shows that he considered that 'negroes in gas masks' could only be effective if led by white officers, an interesting insight into his views on black mental potential.
155   Haldane, *Heredity and Politics*, p. 149.
156   Haldane MSS, MS20594, 'Human Biology and Politics', 1934.
157   Huxley and Haddon, *We Europeans*, p. 280.
158   Arthur Keith MSS, KRP V, Huxley Memorial Lecture, 1928.
159   Keith MSS, KRP V, Huxley Memorial Lecture, 1928.
160   Keith MSS, KRP V, Huxley Memorial Lecture, 1928. Also see A. Keith, 'The Evolution of the Human Races', *Journal of the Royal Anthropological Institute of Great Britain and Ireland* (1925), 58, 306–21, 319.
161   A. Keith, *The Place of Prejudice in Modern Civilisation: (Prejudice and Politics)* (London: Williams and Norgate, 1931), p. 53.
162   Keith MSS, KRP 6, 'An Address to the National Union of Students', 5/4/32.
163   Keith MSS, KRP 6, 'Race and Propaganda': the Bromley Lecture, 1941.
164   Keith MSS, KRP 6, 'Race and Propaganda': the Bromley Lecture, 1941.
165   The racial conservatism of Gates and Keith was fuelled, like Nazism, by conspiratorial anti-Semitism. In this period of racial rethinking, Gates and Keith felt that Jews were responsible for trying to destroy the science of race for their own benefit. As Gates put it in his 1935 notes on the work of American psychologist Otto Klineberg: 'all Jews... [try]...in every way to minimise race differences' (Gates MSS, File 11/41, 1935). The significance of anti-Semitic beliefs in the science of Gates and Keith is discussed at length below in Chapters 3 and 4.
166   See Gates MSS, 11/48. In Gates's thinking, Britons could mix with other Northern European peoples even with a benevolent effect. Also see R. Gates, *Heredity and Politics* (London: Constable, 1923), p. 223.

167 J. Gregory, *The Menace of Colour* (London: Seeley Service, 1925), pp. 323–7.
168 Gregory, *The Menace of Colour*, pp. 323–7.
169 C. Armstrong, *The Survival of the Unfittest* (London: C.W. Daniel, 1927), p. 9.
170 See F. Crew, *Organic Inheritance in Man* (London: Oliver and Boyd, 1927), p. 105. The Eugenics Society continued periodically to voice these concerns into the 1930s through its publication, *Eugenics Review* (see K. Aikman, 'Race Mixture' and A. Ludovici, 'Eugenics and Consanguineous Marriages', *Eugenics Review* (1933), 25: 3, 151–66.
171 Huxley said that Gates had made 'a grave error' in suggesting that the major races 'should be regarded as true species'. Huxley, 'Galton Lecture', p. 17.
172 Huxley MSS, Box 12, Gates to Huxley, 22/2/37.
173 For example, Barkan has highlighted the small scientific difference between Gates and *We Europeans* in *The Retreat of Scientific Racism*, p. 307.
174 Smith, 'Biology and Values', p. 212. Also see Jones, *Science, Politics and the Cold War*, pp. 12–13.
175 See P. Collins, 'The British Association as Public Apologist for Science, 1919–1946', in R. MacLeod and P. Collins (eds), *The Parliament of Science: the British Association for the Advancement of Science, 1831–1981* (Northwood: Science reviews, 1981), W. McGucken, 'The Social Relations of Science: The British Association for the Advancement of Science, 1931–1940', *Proclamations of the American Philosophical Society*, cxiii, 1994. For a contemporary analysis of the changing relationship between science and society see J.G. Crowther, *The Social Relations of Science* (London: Macmillan, 1941).
176 See H.G. Wells, G.P. Wells and J. Huxley, *The Science of Life: a Summary of Contemporary Scientific Knowledge about Life and its Possibilities*, 3 vols (London: Amalgamated Press, 1929–30) and L. Hogben, *Science for the Citizen: a Self Educator Based on the Social Background of Scientific Discovery* (London: Allen and Unwin, 1938). Other notable publications of this kind included Hobgen's *Mathematics for the Million: a Popular Self Educator* (London: Allen and Unwin, 1936) and E. Andrade and J. Huxley, *Simple Science* (Oxford: Basil Blackwood, 1934).
177 Hogben, *Science and the Citizen*, p. 9.
178 See Kohn, *A Reason for Everything*, p. 177 and Werskey, *The Visible College*, pp. 173–5.
179 Werskey, *The Visible College*, p. 175.
180 Haldane MSS, 20537, 'The Scientific Point of View', *The Realist*, July 1929.
181 Hogben, *Science for the Citizen*, p. 1086. Hogben also took forward this agenda in *Dangerous Thoughts*.
182 Hogben, *Science for the Citizen*, p. 1089.
183 See D. Kevles, 'Huxley and the Popularization of Science' in Waters and Van Helden (eds), *Julian Huxley*, pp. 238–51, R. Clark, *The Huxleys* (London: Cox and Wyman, 1968), pp. 179–209 and J. Baker, *Julian Huxley: Scientist and World Citizen 1887–1975* (Paris: UNESCO, 1978), pp. 11–13.

184 Huxley MSS, Box 68, 'Science and Synthesis'.
185 Huxley MSS, Box 13, Fleure to Huxley, 7/3/38. Singer to Huxley, 26/3/38.
186 Mazumdar, *Eugenics*, p. 7.
187 Blacker, 'Galton Lecture', p. 27.
188 Mazumdar has noted that the Eugenics Society was invited to give evidence to Lord Brock's Departmental Committee on Sterilization which reported in 1934 and was 'well represented' on the committee which developed the 1929 Wood Report on 'the incidence of mental deficiency in the population'. See Mazumdar, *Eugenics*, pp. 197–209.
189 Mazumdar, *Eugenics*, p. 7.
190 See Stone, 'Race in British Eugenics', p. 416.
191 Smith, 'Biology and Values in Interwar Britain', p. 212.
192 See Barkan, *The Retreat of Scientific Racism*, and Stepan, *The Idea of Race in Science*.
193 L. London, *Whitehall and the Jews, 1933–48: British Immigration Policy, Jewish Refugees and the Holocaust* (Cambridge: Cambridge University Press, 2000), p. 15. For analysis of the growth of the Jewish community in Britain in this period see T. Endelman, *The Jews of Britain 1656–2000* (Berkeley and London: University of California Press, 2002), and L. Gartner, *History of the Jews in Modern Times* (Oxford: Oxford University Press, 2001).
194 Tony Kushner has highlighted the eclectic roots of popular thinking on Jews in T. Kushner, *Remembering Refugees: Then and Now* (Manchester and New York: Manchester University Press, 2006), p. 108.
195 See B. Cheyette, *Constructions of 'the Jew' in British Literature and Society: Racial Representations 1875–1945* (Cambridge: Cambridge University Press, 1993), pp. 1–13. For broader analysis of the inconstant nature of discourse see M. Foucault, *The History of Sexuality*, Volume 1, *An Introduction* (London: Allen Lane, 1979), p. 33.
196 A. White, *The Modern Jew* (London: Heinemann, 1899), p. 176.
197 White, *The Modern Jew*, p. 79.
198 These characterisations were closely linked to the well-trodden racial analysis which held that Jewish men were more feminine in terms of their reaction to physical danger than British men were. See S. Gilman, *The Jew's Body* (London: Routledge, 1991), p. 63.
199 A. White, *Efficiency and Empire* (London: Methuen, 1901), pp. 79–80.
200 See C. Holmes, 'Public Opinion in England and the Jews 1914–18', *Michael 10* (1986), 76–95 and S. Kadish, *Bolsheviks and British Jews: the Anglo-Jewish Community and the Russian Revolution* (London: Frank Cass, 1992), pp. 10–36.
201 Whilst this figure does not take account of the disproportionately young age of the British Jewish community, it does highlight a substantial Jewish contribution to the war effort. See Kadish, *Bolsheviks and British Jews*, p. 51. Also see M. Levene, 'Going Against the Grain: Two Jewish Memoirs of War and Anti-War 1914–18', *Jewish Culture and History* (1999), 2: 2, 66–95 and D. Cesarani, 'An Embattled Minority: the Jews in Britain during the First World War', *Immigrants and Minorities* (1989), 8: 1–2, 61–81.
202 Moul and Pearson, 'Alien Immigration', p. 199.
203 G. Mudge, 'The Menace to the English Race and to its Traditions of Present Day Immigration and Emigration', *Eugenics Review* (1920), 11: 4, 202–12.

204 For analysis of this kind of prejudice against distinctly Jewish troops in the First World War see M. Watts, *The Jewish Legion and the First World War* (Basingstoke: Palgrave, 2004), pp. 160–81.
205 Parkes Weber MSS, Wellcome Library, London, PP/FPW/B163/1, 'Possible Pitfalls in Life Assurance Examination and Remarks on Malingering' (for the Assurance Medical Society), 1918.
206 Parkes Weber MSS, PP/FPW/163/3. Notes on a patient with shell shock, 1/3/18.
207 Holmes, 'Public Opinion', p. 103.
208 *Jewish Chronicle*, 21/2/19.
209 For further analysis see C. Holmes, *A Tolerant Country? Immigrants, Refugees and Minorities in Britain* (London: Faber and Faber, 1991), p. 27.
210 It is significant that this law was carried through Parliament by Herbert Samuel. The role of Samuel as a British Jew in creating the legislation serves as an indication that the established British Jewish community was both wary of allegations of Jewish 'draft-dodging' and affected by the prevalent racialised thinking concerning new Jewish Russian immigrants. For analysis, see Kadish, *Bolsheviks and British Jews*, pp. 51–69.
211 The attitude of the British government and the dilemma faced by Jewry has been well expressed by Levene. He has labelled the Anglo-Russian Military Service Agreement as a governmental 'not very subtle way of saying good riddance' to Jewish immigrants. Levene has described the dilemma facing Jewry (being compelled to fight for the Russians or the British) as being caught 'between the devil and the deep blue sea' in Levene, 'Going Against the Grain', p. 73. Author, Louis Golding, highlighted the plight of refugee Jews and their reasons for wanting to avoid military service in L. Golding, *Magnolia Street* (St Albans: Gainsborough, 1932), pp. 284–5.
212 Armstrong, *The Survival of the Unfittest*, p. 92. See Parkes Weber MSS, PP/FPW/C10, and Gates, *Heredity and Eugenics*, p. 232.
213 *Hansard*, Vol: 114, Col: 2767, 15/4/19.
214 *Hansard*, Vol: 114, Col: 2799/2801, 15/4/19.
215 *Hansard*, Vol: 114, Col: 2777, 15/4/19.
216 For analysis see Gainer, *Alien Invasion*, pp. 144–98.
217 *Hansard*, Vol: 145, Col: 711, 2/5/05.
218 *Hansard*, Vol: 145, Col: 722, 2/5/05. Also see W. Evans-Gordon, *The Alien Immigrant* (London: William Heinemann, 1903).
219 Parkes Weber MSS, PP/FPW/C10, Weber to Hirschkopf, 9/8/43.
220 Gates MSS, Section 11/5, File: Jews (Undated).
221 A. Memmi, *Portrait of a Jew* (London: Eyre and Spottiswoode, 1963), p. 170.
222 See Kushner, *The Persistence of Prejudice*, pp. 78–106 and T. Kushner, *The Holocaust and the Liberal Imagination: a Social and Cultural History* (Oxford: Blackwell, 1994), p. 273.
223 Wasserstein has argued: 'There was a considerable degree of public sympathy in Britain for the persecuted Jews fleeing the Reich' in B. Wasserstein, *Britain and the Jews of Europe* (Oxford: Oxford University Press, 1988), p. 9.
224 See Kushner, *The Holocaust and the Liberal Imagination*, p. 49. Kushner records that between 'Kristallnacht' and the beginning of the war, Britain gave refuge to 40 per cent of the Jews who managed to escape from Nazi-controlled Europe, p. 51. Also see Holmes, *Anti-Semitism*, p. 218.

225 For example, see the diary of Conservative MP, Henry 'Chips' Channon, who believed that Chamberlain had to (and would) stand firm against the machinations of international Jewish conspiracy: 'The bars and lobbies of the League's buildings are full of Russians and Jews who intrigue with and dominate the press, and spend their time spreading rumours of approaching war, but I don't believe them. Not with Neville at the helm' in R. James (ed.), *Chips: the Diary of Sir Henry Channon*, 12/9/38 (London: Weidenfeld and Nicholson, 1967), p. 164.

226 J. Margach, *The Anatomy of Power: an Enquiry into the Personality of Leadership* (London: W.H. Allen, 1979), p. 53. The Prime Minister notoriously informed his sister on the subject of Jews: 'I don't care about them myself.' See N. Crowson, *Facing Fascism: the Conservative Party and the European Dictators 1935–40* (London: Routledge, 1997), p. 31.

227 See for example the anti-Semitic, pro-German analysis within the well-connected *Truth* magazine, analysed in R. Cockett, *Twilight of Truth: Chamberlain, Appeasement and the Manipulation of the Press* (London: Weidenfeld and Nicholson, 1989), and T. Kushner, 'Clubland, Cricket Tests and Alien Internment 1939–1940', in Kushner and Cesarani (eds), *The Internment of Aliens*, pp. 81–97. *Truth* firmly asserted the idea that war with Germany was unnecessary and was only a worthwhile venture for international Jewry who were (as inherent criminal conspirators) trying to manipulate Britain into joining the struggle. A *Truth* editorial argued, in July 1939, that 'if we set aside the ideal passions of Mr Gollancz and his tribe in the tents of Bloomsbury, the truth is that no appreciable section of British public opinion desires to re-conquer Berlin for the Jews', Cockett, *Twilight of Truth*, p. 163. Ernst Tennant of the Anglo-German Fellowship expressed similar views on the Nazi persecution of Jews. Whilst noting that Hitler's extremism with regard to Jews had been a mistake, Tennant commented: 'That he was justified in reducing Jewish control in certain trades and professions even the best Jews in Germany themselves admit'. Mount Temple MSS, Hartley Archive, University of Southampton, BR81/13, 5/9/34.

228 See Kushner, 'Clubland, Cricket Tests and Alien Internment', pp. 89–94 and Holmes, *A Tolerant Country?*, p. 37.

229 London, *Whitehall and the Jews*, p. 277.

230 London, *Whitehall and the Jews*, p. 276. Hankey was a member of the Constitutional Research Association (CRA). The CRA was an anti-socialist, anti-Semitic and pro-Nazi organisation which offered outspoken opposition to the Nuremberg trials. See G. Macklin, *Very Deeply Dyed in Black: Sir Oswald Mosley and the Resurrection of British Fascism after 1945* (London and New York: I.B. Tauris, 2007), pp. 126–7.

231 PRO, HO 213/1772, Memorandum by A.J. Eagleston for Aliens Branch, 'The Control of Alien Immigration down to 1938'.

232 Eagleston, 'Alien Immigration'.

233 See Kushner and Knox, *Refugees in an Age of Genocide*, pp. 78–100. Also see Eagleston's report on 'Alien Immigration'.

234 Eagleston, 'Alien Immigration'.

235 Eagleston, 'Alien Immigration'.

236 London, *Whitehall and the Jews*, p. 47.

237 Kushner, *Persistence of Prejudice*, p. 197.

238 London, *Whitehall and the Jews*, p. 121. Also see Kushner, *Remembering Refugees*, pp. 141–80.
239 Haldane MSS, 20547, 'Scientific Research', 1934. Haldane noted: 'In most branches of science Britain and the United States are now leading. Germany has dropped out of the race since half her scientific elite, not all Jews, lost their positions and probably ties for third place with Japan.'
240 Haldane MSS, 20594, 'Human Biology and Politics', 1934.
241 L. Golding, *The Jewish Problem* (London: Penguin, 1938), p. 173.

3 The Challenge of War: the 1940s

1 See Jones, *Science, Politics and the Cold War*, p. 13.
2 For example, C.P. Blacker rejoined the Coldstream Guards regiment, in which he had served in the First World War.
3 Werskey, *Visible College*, p. 265. Also see G.P. Wells, 'Lancelot Thomas Hogben', *Biographical Memoirs of Fellows of the Royal Society* (1978), 24, 183–221. Werskey also documents other scientific war roles, pp. 263–6.
4 Kohn, *A Reason for Everything*, p. 177.
5 Haldane wrote at length about the dangers of new weaponry in the interwar period. See especially, *Callinicus* where Haldane argued that the terrible potential of new artillery made chemical warfare a more humane military option. See J.B.S. Haldane, *Callinicus: a Defence of Chemical Warfare* (London: Kegan Paul, 1925).
6 Huxley MSS, Box 15, Hogben to Huxley, 30/9/40.
7 Werskey cites the importance of the anonymously published *Science in War*, which he documents as a product of Solly Zuckerman's 'Tots and Quots' dining club. See Werskey, *The Visible College*, p. 263.
8 J.B.S. Haldane, *Science Advances* (London: Allen and Unwin, 1947), pp. 235–6. Clark recalls that Haldane felt that scientists were being underused in the war. See R. Clark, *The Life and Work of JBS Haldane* (London: Hodder and Stoughton, 1968), pp. 134–5.
9 J. Huxley, 'Men of Science and the War', *Nature* (1940), 146: 3691, 107–8.
10 J. Huxley, 'Scientists in Uniform: England', *The New Republic*, 28/4/41, 590–1, 590.
11 Joseph Needham shared similar views. He argued: 'should the Nazis be victorious, science in Europe may disappear for several generations, and all social progress with it' in J. Needham, *The Nazi Attack on International Science* (London: Watts and Co, 1941), p. 47.
12 Hogben, *Dangerous Thoughts*, p. 50.
13 See Kohn, *A Reason for Everything*, p. 177.
14 Werskey, *The Visible College*, pp. 173–4. Werskey argues that Haldane was the most famous of those scientists in the 'Visible College' in the 1930s, p. 174.
15 For Huxley's work on the *Brains Trust* see Clark, *The Huxleys*, pp. 278–80 and D. Lamahieu, 'The Ambiguity of Popularisation' in Waters and Van Helden, *Julian Huxley*, pp. 255–6.
16 For details see S. Nicholas, *The BBC, British Morale, and the Home Front War Effort 1939–45* (Oxford: Oxford University Press, 1992) and S. Nicholas, in N. Hayes and J. Hill (eds), *'Millions like us?': British Culture in the Second World War* (Liverpool: Liverpool University Press, 1999).

17 P. Black, 'When I was in Patagonia', *Daily Mail*, 28/11/58.
18 *The Listener*, 5/5/60.
19 S. Nicholas, *The Echo of War: Home Front Propaganda and the Wartime BBC 1939–45* (Manchester and New York: Manchester University Press, 1996), p. 140.
20 The *Brains Trust* itself steered clear of controversial topics according to Nicholas, *The Echo*, p. 140.
21 R. Gregory, *Science in Chains* (London: Macmillan, 1941), p. 3.
22 This question was posed by Watson Davis, editor of *Science Service*. See 'Men and Mice at Edinburgh: Reports from the Genetics Congress', *Journal of Heredity* (1939), 30: 9, 371–4.
23 'Men and Mice', p. 372.
24 'Men and Mice', p. 372.
25 'Men and Mice', p. 372. This view resonated with the thinking of other progressive signatories. For example, Lancelot Hogben wrote in *The Retreat from Reason* (London: Watts and Co, 1936): 'the racialist doctrine has no serious title to be accepted as a biological interpretation of history, until the ecological factors in cultural differentiation have been explored exhaustively', p. 42.
26 Barkan, *The Retreat of Scientific Racism*, p. 280.
27 Werskey, *Visible College*, p. 267.
28 'Men and Mice', p. 372.
29 Jones, *Science, Politics and the Cold War*, pp. 3–4.
30 'Men and Mice', p. 372.
31 For the origins of Huxley's involvement with UNESCO see J. Huxley, *Memories II* (London: Allen and Unwin, 1973), pp. 30–6.
32 Needham, 'The Nazi Attack', p. 14. At this time Needham was a Reader in Biochemistry at the University of Cambridge.
33 Huxley MSS, Box 13, Letter to Dr Waldemar Kaempffert, science editor of the *New York Times*, 28/4/40.
34 Haldane MSS, MS20604, 'Essay on Race', C1945–46.
35 Huxley MSS, Letter to Kaempffert, 28/4/40.
36 J. Haldane, *Science and Everyday Life*, p. 180.
37 Haldane, *Science and Everyday Life*, p. 179.
38 Huxley argued that there were more 'flat heads' in Switzerland and Czechoslovakia than in Germany and dismissed Twyman's argument that 'flat heads' were stupid noting that Galileo was a 'flat head'. Huxley MSS, Box 15, Huxley to Twyman, 8/10/40.
39 Huxley MSS, Box 15, Huxley to Maurice Tweedale, 4/6/41.
40 Haldane MSS, 20604, 'Paper on Race', 1945–6.
41 Haldane MSS, 20549. 'Our Men and Women', *Reynolds News*, November 1950.
42 J. Huxley, *The Uniqueness of Man* (London: Chatto and Windus, 1941), p. 49.
43 J.B.S. Haldane, *Keeping Cool and Other Essays* (London: Chatto and Windus, 1940), pp. 34–5.
44 Haldane MSS, 20562, 'What is Race?', Handwritten notes (undated). Also see Haldane, *Keeping Cool*, pp. 35–6.
45 Huxley, *The Uniqueness of Man*, p. 53.
46 Haldane, *Keeping Cool*, p. 35.

47 Haldane, 'Our Men and Women'.
48 Hogben, *Dangerous Thoughts*, p. 47.
49 Hogben, *Dangerous Thoughts*, p. 45.
50 Needham, *The Nazi Attack*, p. 15.
51 Needham, *The Nazi Attack*, p. 39.
52 See Jones, *Science, Politics and the Cold War*, p. 75 and R. Darnell, *And Along Came Boas: Continuity and Revolution in American Anthropology* (Amsterdam: J. Benjamin, 1998).
53 See W. Jackson, *Gunnar Myrdal and America's Conscience* (Chapel Hill: University of North Carolina Press, 1990), D. Southern, *Gunnar Myrdal and Black–White Relations* (Baton Rouge: Louisiana State University Press, 1987) and King, *Race Culture and the Intellectuals*, pp. 21–48.
54 Jackson, *Gunnar Myrdal*, pp. xvi–xvii.
55 G. Myrdal, *An American Dilemma: the Negro Problem and Modern Democracy* (New York and London: Harper, 1944).
56 King, *Race, Culture and the Intellectuals*, p. 22.
57 Jackson, *Gunnar Myrdal*, pp. 274–5 and King, *Race, Culture and the Intellectuals*, pp. 22–3. For broader analysis of the use of science in legal challenges to segregation see J. Jackson, 'The Scientific Attack on Brown v. Board of Education, 1954–1964', *American Psychologist* (2004), 59: 6, 530–7.
58 O. Klineberg (ed.), *Characteristics of the American Negro* (New York: Harper, 1944). Also see R. Benedict, *Race, Science and Politics* (New York: New Age Books, 1940) and A. Montagu, *Man's Most Dangerous Myth: the Fallacy of Race* (New York: Columbia University Press, 1945). For British contributions to this new anthropology see D. Stone, 'Nazism as Modern Magic: Bronislaw Malinowski's Political Anthropology', *History and Anthropology* (2003), 14: 3, 203–18.
59 Needham, *The Nazi Attack*, p. 44.
60 Needham, *The Nazi Attack*, p. 26.
61 Huxley in I. Zollschan, *Racialism Against Civilisation* (London: New Europe, 1942), p. 8.
62 Little published his first major piece of research on this topic in 1948. See K. Little, *Negroes in Britain: a Study of Race Relations in English Society* (London and Boston: Routledge and Kegan Paul, 1948). In the war years he was already publishing on the topic. See K. Little, 'The Study of Racial Mixture in the British Commonwealth: Some Anthropological Preliminaries', *The Eugenics Review* (1941), 32: 4, 114–20 and 'Racial Mixture in Great Britain: Some Anthropological Characteristics of the Anglo-Negroid Cross', *Eugenics Review* (1942), 33: 4, 112–20.
63 Huxley MSS, Box 65, 'Black Man's Country', February 1945.
64 See R. Clark, *The Rise of the Boffins* (London: Phoenix, 1962).
65 Werskey, *Visible College*, pp. 271–2.
66 Huxley MSS, Box 13, Huxley to Prof. Quincy Wright, University of Chicago, 6/12/39. For Needham see Werskey, *Visible College*, pp. 265–6.
67 Blacker MSS, Box 2, A4/5, Carr Saunders to Blacker, 14/11/39.
68 The true goal of Huxley's trip was secret enough to have been kept from his employers at the Zoological Society who were exasperated by his unexplained extended leave of absence. (See Huxley MSS, Box 13, Kenneth Clark to Huxley, 13/3/40.) Despite its unofficial nature it was

sufficiently controversial to generate extremely hostile correspondence from the American public. See Huxley MSS, Box 13.
69 Huxley MSS, Box 13, Bevan to Huxley, 24/4/40.
70 A.A. Milne wrote 'War Without Honour', E.M. Forster, 'Nordic Twilight' and Joad wrote 'For Civilisation'. Another scientific contribution was made by eugenicist R.A. Gregory who wrote 'Science in Chains'. Gregory made his commitment to the anti-Nazi cause clear in correspondence with Singer. He wrote: 'If there ever was a war between good and evil, this is one and it ought to be proclaimed from the house tops.' Singer MSS, CJS A7-1, 14/9/39.
71 Huxley MSS, Box 13, Bevan to Huxley, 24/4/40.
72 Huxley, *Argument of Blood* (Inside cover).
73 Huxley MSS, Box 13, Huxley to Bevan, 27/4/40.
74 For Singer's agreement to meet Greene at the Ministry of Information see Huxley MSS, Box 13, Singer to Huxley, 15/5/40. The pamphlet appeared as J. Huxley, *Argument of Blood* (London: Macmillan, 1941).
75 Huxley, *Argument of Blood*, pp. 4–5.
76 Huxley, *Argument of Blood*, pp. 40–4.
77 This text is unusual in its frequent references to Jewish suffering under the Nazis. British propaganda in this period was consciously focused by the government away from presenting Jews as the primary victims of Nazism. It seems that neither Singer nor Huxley were prepared to subscribe to this tactic.
78 Huxley MSS, Box 13, Huxley to Singer, 6/8/40.
79 Huxley MSS, Box 13, Huxley to Greene, 7/9/40.
80 Huxley MSS, Box 13, Singer to Huxley, 6/8/40. Singer badgered Huxley for months over delays to the publication of the pamphlet. See Huxley MSS, Box 13, Singer to Huxley, 10/12/40. Singer writes again, Box 15, Singer to Huxley, 12/2/41.
81 In correspondence with Huxley, Singer refers to the work as 'your pamphlet'. Huxley MSS, Box 13, Singer to Huxley, 10/12/40. Huxley continues to refer to it as 'Singer's pamphlet'. See Huxley MSS, Box 13, Huxley to Greene, 19/9/40.
82 Singer MSS, File B16, Singer to Edgar Ashworth Underwood, 12/10/53.
83 During this conflict, 30,000 Germans living in Britain were interned and 10,000 more were forced to leave the country. For details see P. Panayi, *The Enemy in Our Midst: Germans in Britain during the First World War* (Oxford: Berg, 1991) and P. and L. Gillman, *Collar the Lot! – How Britain Interned and Expelled its Wartime Refugees* (London: Quartet Books, 1980), pp. 8–21.
84 Approaching the Second World War, a widespread feeling existed that internees in future required more careful vetting to prevent unnecessary hardships (a government committee was established in March 1938 to work on the future classification of 'aliens'). Additionally, First World War experience led to a governmental preference, though not in the event possible to achieve, that no substantial numbers of enemy 'aliens' should remain in Britain during any future conflict. See R. Stent, *A Bespattered Page? The Internment of 'His Majesty's Most Loyal Enemy Aliens'* (London: André Deutsch, 1980), pp. 18–22. Also see Cesarani, 'An Alien Concept?', pp. 25–52.

85 See Stent, *A Bespattered Page*, pp. 23–9, Kushner, *The Persistence of Prejudice*, p. 144 and Gillman and Gillman, *Collar the Lot!*, pp. 100–10.
86 For details see F. Lafitte, *The Internment of Aliens* (Harmondsworth: Penguin, 1940), pp. 62–5 and Stent, *A Bespattered Page?*, pp. 30–42.
87 Stent, *A Bespattered Page?*, p. 37.
88 Whilst this group contained some overt supporters of Nazism, other internees were refugee Jews and other anti-Nazi unfortunates who had been given an 'A' or 'B' risk classification in error. In fact, some scholars have argued that the majority of category 'A' internees were anti-Nazis. See Stent, *A Bespattered Page?*, pp. 34–41 and M. Kochan, *Britain's Internees in the Second World War* (London and Basingstoke: Macmillan, 1983), pp. 9–13.
89 Some analysts have cited the primacy of elements of the press in creating an anti-'alien' frenzy. See T. Kushner, '"Beyond the Pale", British Reaction to Nazi Anti-Semitism 1933–9', in T. Kushner and K. Lunn (eds), *The Politics of Marginality: Race, the Radical Right and Minorities in Twentieth Century Britain* (London: Frank Cass, 1990), pp. 143–60 and Lafitte, *The Internment of Aliens*, pp. 167–71. Stent has cited *Daily Mail* reporter, G. Ward Price, as playing a unique role in the creation of pro-internment feeling. See Stent, *A Bespattered Page?*, p. 47. There is general agreement that various individuals in government circles played a key role. Most notably, British Ambassador to the Netherlands, Neville Bland, produced a hysterical report after the Nazi invasion entitled 'The Fifth Column Menace' alleging the complicity of 'aliens' in the fall of Holland. See PRO, FO371/25189/462, Bland to Foreign Office, 14/5/40. A similar report was produced by the Joint Chiefs of Staff entitled 'British Strategy in a Certain Eventuality'. For analysis, see Gillman and Gillman, *Collar the Lot!*, pp. 101–2 and Lafitte, *The Internment of Aliens*, pp. 172–3.
90 Some analysts have highlighted the role of the Home Defence (Security) Committee under Viscount Swinton of Masham. Neither the exact membership details or terms of reference of this committee are known but Gillman and Gillman have highlighted its influence in *Collar the Lot!*, pp. 144–5. Also see Stent, *A Bespattered Page?*, pp. 182–3.
91 For details see Lafitte, *The Internment of Aliens*, pp. 75–91.
92 Burletson has questioned whether the decision to move 'high risk aliens' abroad was also taken by the Swinton Committee. See L. Burletson, 'The State, Internment and Public Criticism in the Second World War', in Cesarani and Kushner, *The Internment of Aliens*, pp. 115–16.
93 For details see Stent, *A Bespattered Page?*, pp. 100–9.
94 Stent, *A Bespattered Page?*, pp. 114–33.
95 See Kushner, *The Persistence of Prejudice*, p. 147.
96 See Stent, *A Bespattered Page?*, pp. 240–51.
97 See Cesarani, 'An Alien Concept?', pp. 33–5 and Lafitte, *The Internment of Aliens*, pp. 161–91.
98 Kushner, *The Persistence of Prejudice*, p. 147.
99 For accounts of governmental responses to 'fifth column' fears see Kushner, *The Persistence of Prejudice*, pp. 145–7, Gillman and Gillman, *Collar the Lot!*, pp. 100–45, Lafitte, *The Internment of Aliens*, pp. 172–7 and Stent, *A Bespattered Page?*, pp. 53–9.

100 *Hansard*, Vol: 364, Col: 1539, 12/8/40.
101 PRO, FO371/25189/462, Bland to Foreign Office, 14/5/40. For a history of Jewish refugees in domestic service see T. Kushner, 'Politics and Race, Gender and Class: Refugees, Fascists and Domestic Service in Britain, 1933–40' in Kushner and Lunn, *The Politics of Marginality*, pp. 49–60.
102 Many Jewish refugees found it difficult to understand that they could be perceived as potential allies to Hitler. See Kochan, *Britain's Internees*, p. 34, E. Spier, *The Protecting Power* (London: Skeffington, 1951). Also see Kushner, *The Persistence of Prejudice*, p. 150.
103 *Hansard*, Vol: 365, Col: 1564, 12/8/40.
104 *Hansard*, Vol: 365, Col: 1564, 12/8/40.
105 *Hansard*, Vol: 362, Col: 1220, 11/7/40.
106 The following argument from Winterton reflected many 'benevolent' governmental voices demanding internment: 'I am convinced that cases did occur where Germans said to refugees, "We are going to give you and your people a much worse time, but we will give you the opportunity, if you like, of going to other countries providing you will help us in any way you can."' See *Hansard*, Vol: 365, Col: 1540, 12/8/40.
107 For images of the Jewish coward in the First World War see Holmes, 'Public Opinion in England and the Jews 1914–18', pp. 76–95 or Kadish, *Bolsheviks and British Jews*, pp. 10–36.
108 Kushner has highlighted the racial nature of thinking of this kind: 'On pure statistical grounds there was again no basis for the Jewish war shirker image to come about. To explain its pervasive appeal one has, as usual, to examine the past Jewish stereotype.' See Kushner, *The Persistence of Prejudice*, p. 122.
109 Board of Deputies MSS, London Metropolitan Archive, E3/536, 21/9/43.
110 PRO, FO 371/24100, Reilly to Halifax, 29/7/39.
111 Board of Deputies MSS, E3/520, 2/4/40.
112 *Hansard*, Vol: 364, Col: 1546, 12/8/40.
113 *Hansard*, Vol: 364, Col: 1579, 12/8/40.
114 Haldane MSS, 20534, Hans Simon to Haldane, 13/5/33.
115 Huxley MSS, Box 13, Woburn House to Huxley regarding a Mr and Mrs Szues, 18/11/39.
116 Huxley MSS, Box 15, Free German League of Culture to Huxley, 28/11/40.
117 For Singer's involvement with the Central Office for Refugees see Singer MSS, File E1.
118 Huxley MSS, Box 13, Singer to Huxley, 24/4/40.
119 Kushner and Knox, *Refugees in an Age of Genocide*, p. 161. For details see N. Bentwich, *The Rescue and Achievement of Refugee Scholars: the Story of Displaced Scholars and Scientists 1933–52* (The Hague: Nijhoff, 1953). For analysis see R. Cooper, *Refugee Scholars: Conversations with Tess Simpson* (Leeds: Moorland, 1992), R. Cooper, *Retrospective Sympathetic Affection: a Tribute to the Academic Community* (Leeds: Moorland, 1996) and London, *Whitehall and the Jews*, pp. 47–50.
120 Parkes was a leading campaigner on behalf of Europe's Jews. See J. Parkes, *An Enemy of the People: AntiSemitism* (Harmondsworth: Penguin, 1945). For analysis see C. Richmond, *Campaigner against AntiSemitism: the Reverend James Parkes, 1896–1981* (London: Vallentine Mitchell, 2005).

121 See Cooper, *Refugee Scholars*, p. 32.
122 Kushner and Knox, *Refugees in an Age of Genocide*, p. 161.
123 Huxley MSS, Box 13, Esther Simpson to Huxley, 28/7/40.
124 Huxley MSS, Box 13, SPSL to Huxley, 23/7/40.
125 Huxley MSS, Box 13, Esther Simpson to Huxley, 27/8/40.
126 Huxley MSS, Box 13, Esther Simpson to Huxley, 27/8/40. Cooper has downplayed the differences between the Royal Society and the government, arguing that within three months all the applicants were released in *Refugee Scholars*, p. 43. However, this timescale corresponds with the ending of internment, making it difficult to assess the impact of the Society's evaluation of internees' scholarship.
127 Huxley MSS, Box 15, SPSL to Huxley, 1/11/40.
128 Huxley MSS, Box 13, Huxley to camp commander and F.A. Newsam, 25/7/40. To Lord Horder, 20/7/40.
129 Huxley MSS, Box 13, Huxley to Horder, 20/7/40 and 26/7/40.
130 Huxley MSS, Box 99, 'The Growth of a Group Mind in Britain under the Influence of War', *The Hibbert Journal*, July 1941, 343–4.
131 Huxley MSS, Box 13, Huxley to Mrs Honigmann, 27/6/40.
132 Letter to the editor of the *Picture Post*, 4/5/40.
133 Huxley MSS, 'The Growth of a Group Mind', p. 344.
134 London is correct to note that the help provided was only to the scientific elite. This may suggest a eugenic agenda, as scientists wanted only the cream of Germany's academics, not to help every German in need. See London, *Whitehall and the Jews*, p. 48.
135 See G. Schaffer, 'Re-Thinking the History of Blame: Britain and Minorities during the Second World War', *National Identities* (2006), 8: 4, 401–20.
136 For a sophisticated analysis of black US soldier presence in Britain see C. Thorne, 'Britain and the Black GIs: Race Issues and Anglo-American Relations in 1942', in C. Thorne, *Border Crossings: Studies in International History* (Oxford: Basil Blackwood, 1988), pp. 259–74, G. Smith, *When Jim Crow Met John Bull* (London: I.B. Tauris, 1987) and D. Reynolds, *Rich Relations: the American Occupation of Britain 1942–5* (London: HarperCollins, 1995), pp. 216–35.
137 M. Sherwood, *Many Struggles: West Indian Workers and Service Personnel in Britain 1939–45* (London: Karia Press, 1985), p. 101.
138 Sherwood, *Many Struggles*, p. 113.
139 Sherwood, *Many Struggles*, pp. 113–16.
140 For example, Reynolds has recorded how police prosecuted mixed race courting couples for damaging crops and by using defence regulations on trespass. Some white women were imprisoned for up to three months as a result of prosecutions of this nature. See Reynolds, *Rich Relations*, p. 229.
141 PRO, PREM 4/26/9, Grigg to Churchill, 21/10/43.
142 PRO, PREM 4/26/9, Grigg Report for Cabinet, December 1942.
143 PRO, PREM 4/26/9, Grigg Report for Cabinet, December 1942.
144 *Huddersfield Daily Examiner*, 20/9/40.
145 A. Richmond, *Colour Prejudice in Britain: a Study of West Indian Workers in Liverpool 1941–51* (London: Routledge and Kegan Paul, 1954), p. 78.
146 PRO, PREM 4/26/9, Marlborough to Churchill (Undated), 1943.
147 PRO, PREM 4/26/9, Grigg to Churchill, 2/12/43.

148 PRO, CO968/17/5, Grigg to Stanley, 23/9/43.
149 In the case of the Honduran Foresters, one day a week at a local surgery was dedicated to treating black troops suffering from VD. See Sherwood, *Many Struggles*: 'An inordinate amount of attention was paid by the Ministry to the incidence of venereal disease among the men', pp. 106–7.
150 Smith, *When Jim Crow Met John Bull*, p. 84.
151 For the government view, see D. Reynolds, 'The Churchill Government and the Black American Troops in Britain during World War Two', *Transactions of the Royal Historical Society* (1985), 5: 35, 113–33.
152 PRO, PREM4/29/6, Report by Grigg for the War Cabinet, September 1942.
153 PRO, CO968/17/5, Grigg to Stanley, 23/9/43.
154 See Reynolds, *Rich Relations*, pp. 216–40 and Smith, *When Jim Crow Met John Bull*, pp. 37–96.
155 PRO, CO937/6/3, Watson to Megson, 2/9/44.
156 PRO, CO968/17/5, War Office Report, 17/12/43.
157 PRO, CO968/17/5, War Office Report, 17/12/43.
158 Also see Sherwood, *Many Struggles*, pp. 28–38.
159 PRO, CO968/17/5, Mayle memorandum, 30/4/43.
160 Colour Restrictions were entrenched in Army Order 89 and Section 95 of the Air Force Act.
161 PRO, CO323/1673, Note by Lees, 14/10/39.
162 Sherwood, *Many Struggles*, pp. 3–25.
163 Sherwood, *Many Struggles*, p. 11.
164 Sherwood, *Many Struggles*: 'What emerges from the remaining files is a picture of racism and racial prejudice throughout the ranks of British society: black men were not wanted in Britain even in times of crisis', p. 51. Also see Sherwood, pp. 124–6.
165 See Dubow, *Scientific Racism*, pp. 191–4.
166 Wells, 'Biographical Memoirs', p. 196. Also see Werskey, *The Visible College*, pp. 106–7.
167 Hogben MSS, File A21 (undated).
168 Hogben, *Dangerous Thoughts*, p. 47.
169 Haldane MSS, 20561, Folio 163, notes (undated). For attack on colour bar see Haldane, *Heredity and Politics*, p. 163.
170 Haldane, *Science and Everyday Life*, p. 178.
171 Huxley, *Africa View*, p. 393.
172 Huxley, *Africa View*, p. 443.
173 See J. Huxley, 'The Future of Colonies', *The Tribune*, 19/1/40.
174 Huxley MSS, Box 11, Evans to Huxley, 1/8/35.
175 For example see Gregory, *The Menace of Colour*, on the impossibility of white settlement in tropical climates. Gregory concluded: 'These simple facts are so impressive that those who hold that the white man can colonise in the tropics have to face a strong prepossession to the contrary', p. 29.
176 Haldane, *Keeping Cool*, p. 36.
177 Haldane, *Heredity and Politics*, p. 169.
178 Haldane MSS, 20594, 'Human Biology and Politics', 1934.
179 Huxley MSS, Box 15, Huxley to Halifax, 25/10/40.
180 Huxley MSS, Box 15, Halifax to Huxley, 14/11/40.
181 Huxley MSS, Box 15, Huxley to Halifax, 25/10/40.

182 Huxley MSS, Box 15, Huxley to Routh, 11/10/40.
183 Huxley MSS, Box 15, Routh to Huxley, 21/10/40.
184 Huxley MSS, Box 16, Eden's acknowledgement of receiving *Reconstruction and Peace*, 27/10/42. Stanley to Huxley, 2/2/43.
185 Huxley's role in the commission is documented in the Huxley MSS, Box 17, Notes on the commission, 25/9/44.
186 J. Huxley, 'The Future of Colonies', *The Tribune*, 19/1/40.
187 J. Huxley, 'The Social Experiment – White and Black Must Work Together', *London Calling*, 210, 1943.
188 Huxley MSS, Box 17, Boycott to Huxley, 1/11/44. Creech Jones replied that he had found the letter 'most interesting' and had sent it on to 'five selected commissioners', 16/11/44.
189 A. Keith, *Essays on Human Evolution* (London: Watts and Co, 1946), pp. 91 and 11.
190 Keith, *Essays on Human Evolution*, p. 9.
191 Keith, *Essays on Human Evolution*, p. 9 and p. 91.
192 Keith MSS, File XAB, 'Race and Propaganda', 1941.
193 Keith, *Essays on Human Evolution*, p. 175.
194 R. Gates, *Human Genetics* (New York: Macmillan, 1946), two volumes, pp. 45, 1160 and 1243.
195 Gates MSS, File 11/48, Comments on A. Montagu, 'The Creative Power of Ethnic Mixture', 1942. Montagu's essay was later included in his *Man's Most Dangerous Myth*.
196 Gates MSS, File 11/41, Gates on Klineberg, *Characteristics of the American Negro*. In describing the 'Myrdal Scheme' Gates was clearly referring to the research of Gunnar Myrdal, specifically to *An American Dilemma*, which Gates believed was written deliberately to erode the racial divisions of the USA.
197 Blacker MSS, Box 3, A4/10, The Galton Lecture, 16/2/45.
198 Blacker MSS, Box 3, A4/10, The Galton Lecture, 16/2/45.
199 C.P. Blacker, 'Galton's Views on Race', *Eugenics Review* (1951), 43: 1, 22.
200 C.P. Blacker, '"Eugenic" Experiments Conducted by the Nazis on Human Subjects', *Eugenics Review* (1952), 44: 1, 9–19.
201 Soloway dates this shift in focus as occurring as early as the 1930s in Soloway, *Demography and Degeneration*, p. 354.
202 Blacker MSS, Box 2, Blacker to Baker, 6/5/69.
203 For details of the origins of the commission see Grebenik, 'Demographic Research', pp. 14–15, Soloway, *Demography and Degeneration*, pp. 312–13 and C. Langord, 'The Eugenics Society and the Development of Demography in Britain: the International Population Union, the British Population Society and the Population Investigation Committee' in Peel, *Essays in the History of Eugenics*, pp. 81–111.
204 Blacker MSS, Box 3, A4/10, The Galton Lecture, 16/2/45.
205 Grebenik, 'Demographic Research', p. 15.
206 Blacker attended the commission on behalf of the Society along with Aubrey Lewis and J.A. Frazer Roberts. Soloway has argued that eugenicists were 'strongly represented' within the Commission's committees and retained 'substantial influence' in Soloway, *Demography and Degeneration*, pp. 338–9.

207 Blacker MSS, Box 23, 'The Royal Commission and the Society's Aims', 22/7/49.
208 Blacker MSS, Box 23, 'The Royal Commission and the Society's Aims', 22/7/49.
209 See Soloway, 'From Mainline to Reform Eugenics' in Peel, *Essays in the History of Eugenics*, p. 76.
210 The Report of the Royal Commission on Population, Cmd 7695, 1949, p. 154.
211 The Report recorded: 'It is clearly undesirable for the welfare and cultural standards of the nation that our social arrangements should be such as to induce those in the higher income groups and the better educated and more intelligent within each income group to keep their families not only below replacement level but below the level of others.' The Report of the Royal Commission, p. 156. For analysis see Soloway, *Demography and Degeneration*, p. 348.
212 The Report of the Royal Commission, pp. 227–31.
213 Soloway, *Demography and Degeneration*, p. 349.
214 Papers of the Royal Commission on Population, Vol: 4, Reports of the Biological and Medical Committee (London: His Majesty's Stationery Office, 1950).
215 Report of the Royal Commission, p. 61.
216 Grebenik, 'Demographic Research', p. 15. Also see Soloway, *Demography and Degeneration*, p. 350.
217 For an analysis of these ongoing concerns see E. Stadulis, 'The Resettlement of Displaced Persons in the United Kingdom', *Population Studies* (1952), 5: 3, 207–37, 208–9.
218 A letter from R.A. Gregory in 1949 suggests that this decision had been made without Gates's consent. See Gates MSS, File 1/12, Gregory to Gates, 19/7/49. 'I am glad to have a letter from you but very sorry to know that you seem to have decided to remain in the USA, because there is no place now for you to work in Great Britain on account of your having been retired from King's College London at the end of the War without consulting you.'
219 This research was published as R. Gates, *Pedigrees of Negro Families* (Philadelphia: Blackiston, 1949). For an analysis of Gates's problems at Howard see Schaffer, 'Scientific Racism Again?', pp. 253–78.
220 For the ideology behind the founding of Howard see R. Logan, *Howard University: the First Hundred Years 1867–1967* (New York: New York University Press, 1969), pp. 18–22.
221 Logan has argued: '...the small number of white students was a matter of their choice and not of university policy' in Logan, *Howard University*, p. 578.
222 Logan, *Howard University*, pp. 361–2.
223 DuBois's project was titled 'A Study of Economic and Social Problems of Negroes'. Work was researching a 'Bibliography on European Colonisation and the Resulting Contacts on Peoples, Races and Cultures'.
224 For analysis of this incident see Jones, *Science, Politics and the Cold War*, pp. 119–21.
225 Gates MSS, Box 1/7, Letter to Dean Price, 17/2/47.

226 As evidence, the letter writers cited Gates's 1944 contribution to the *American Journal of Physical Anthropology* (titled 'Different Species') and his 1934 article in *Population* (titled 'Heredity and Racial and Social Problems').
227 Gates MSS, Box 1/7, Gates to Dean Price, 17/2/47.
228 Gates MSS, Box 1/7, Gates to Dean Price, 22/2/47.
229 Gates MSS, File 11/41, Notes on Otto Klineberg, 1944.
230 Gates MSS, File 1/7, Gates to Martial, 25/4/47.
231 Gates MSS, File 1/7, Gates to Mordecai Johnson, 17/3/47.
232 Gates MSS, Box 1/7, Gates to Dean Price, 17/2/47.
233 Gates MSS, Box 1/7, Gates to Dean Price, 22/2/47.
234 Gates MSS, Box 1/7, Gates to Dean Price, 17/2/47.
235 Gates MSS, Box 1/7, Ellinger to Dean Price, 17/3/47. Jones has recorded that Ellinger was also soon forced out of Howard. See Jones, *Science, Politics and the Cold War*, p. 120.
236 Gates MSS, Box 1/12, Cummins to Gates, 4/4/50.
237 Gates MSS, Box 1/7, Gates to Dean Price, 22/2/47.
238 Gates MSS, Box 1/7, Gates to Dean Price, 22/2/47. It was common for pro-segregation ideologues to claim that they had the best interests of black people at heart. Gates typically commented on one pro-segregation study: 'The author, like Southerners generally, shows more friendliness and real understanding of the Negro than is generally found in the North, where relatively few are present except in the cities.' See Gates MSS, Box 11/102, Review of C. Putnam, *Race and Reason: a Yankee View*, 1961.
239 For details of government thinking on British population trends and manpower needs in this period see Stadulis, 'The Resettlement of Displaced Persons in the United Kingdom', pp. 209–10, D. Cesarani, *Justice Delayed: How Britain Became a Refuge for Nazi War Criminals* (London: Heinemann, 1992), pp. 68–70, J. Isaac, *British Post-War Migration* (Cambridge: Cambridge University Press, 1954), p. 18 and J. Tannahill, *European Volunteer Workers in Britain* (Manchester: Manchester University Press, 1958), p. 114.
240 For analysis of these programmes see D. Kay and R. Miles, *Refugees or Migrant Workers? European Volunteer Workers in Britain 1946–1951* (London: Routledge, 1992), pp. 42–65, Tannahill, *European Volunteer Workers in Britain*, pp. 19–33, Cesarani, *Justice Delayed*, pp. 68–81, K. Paul, *Whitewashing Britain* (Ithaca and London: Cornell University Press, 1997), pp. 64–89 and Holmes, *John Bull's Island*, pp. 210–18.
241 Royal Commission on Population, p. 125.
242 Royal Commission on Population, p. 124.
243 Royal Commission on Population, p. 124.
244 Royal Commission on Population, p. 225.
245 See Isaac, *British Post-War Migration*, p. 232.
246 Royal Commission on Population, p. 124.
247 PRO, FO945/500, Note from Vienna to London Control Office, 31/3/47.
248 PRO, FO945/500. Note by the Allied Commission for Austria on the Recruitment of Displaced Persons for Great Britain, 24/3/47.
249 See Paul, *Whitewashing Britain*, pp. 84–8, Carter, *Realism and Racism*, pp. 113–17, Kay and Miles, *Refugees or Migrant Workers?*, pp. 165–76 and Cesarani, *Justice Delayed*, pp. 70–8.

250 PRO, FO945/500, Note by the Allied Commission for Austria on the Recruitment of Displaced Persons for Great Britain, 24/3/47.
251 Analysts of the EVW programmes have argued that Western European and Baltic people were prioritised as the best potential workers that could be brought to Britain amid the belief that these immigrants could assimilate most successfully with indigenous Britons. See Kay and Miles, *Refugees or Migrant Workers?*, p. 175, Paul, *Whitewashing Britain*, pp. 86–8 and Cesarani, *Justice Delayed*, p. 73.
252 PRO, FO945/501, Winterton to Packenham, April, 1947.
253 PRO, FO371/66709, Wilkinson memorandum, 24/1/47.
254 PRO, FO371/66709. Minutes of the Cabinet Foreign Labour Committee, 14/2/47. Tannahill has noted government concerns that the best workers would be taken by other countries in Tannahill, *European Volunteer Workers in Britain*, p. 42.
255 'Volksdeutsche' was a much contested Nazi term used to describe citizens of non-German countries who were perceived as German by ethnic origin. Many of these people became refugees after German defeat in the war, encountering persecution and hostility in their countries amid allegations of complicity with Nazism.
256 PRO, FO945/500, Conference on the Extension of the 'Westward Ho' Scheme to Austria, 11/4/47.
257 PRO, FO945/501, Chancellor to Winterton (Undated), 1947.
258 PRO, FO945/501, Winterton to Pakenham, April, 1947.
259 See M. Marrus, *The Unwanted: European Refugees in the Twentieth Century* (Oxford and New York: Oxford University Press, 1985), p. 333.
260 Jewish terrorism in Palestine had been muted during the war as the leadership of the *Yishuv* fell in behind the Allies in their fight against Nazism. Terrorism began in earnest after the end of the conflict and was especially pronounced as radical Jewish groups splintered off from the more moderate *Haganah*. For details see M. Gilbert, *Exile and Return: the Emergence of Jewish Statehood* (London: Weidenfeld and Nicolson, 1978), pp. 272–96 and T. Segev, *One Palestine, Complete: Jews and Arabs under the British Mandate* (New York: Metropolitan Books, 2000), pp. 112–50.
261 T. Kushner, 'Anti-Semitism and Austerity: the August 1947 Riots in Britain', in P. Panayi, *Racial Violence in Britain 1840–1950* (Leicester: Leicester University Press, 1993), pp. 152–66. Also see G. Alderman, *The Jewish Community and British Politics* (Oxford: Clarendon, 1983), pp. 128–50, Holmes, *John Bull's Island*, p. 245 and D. Leitch, 'Explosion at the King David Hotel', in M. Sissons and P. French, *The Age of Austerity* (Oxford: Oxford University Press, 1986), pp. 58–85.
262 Kushner, *The Holocaust and the Liberal Imagination*, p. 276.
263 *Hansard*, Vol: 441, Col: 2350, 12/8/47.
264 See Marrus, *The Unwanted*, p. 333.
265 PRO, LAB 8/99, Bevin to Isaacs, 21/2/46.
266 Anglo-Jewish Association MSS, Hartley Archive, University of Southampton, AJ95/148, 'The Jews of Britain' (undated).
267 Kushner, *The Holocaust and the Liberal Imagination*, p. 234.
268 This programme was humanitarian in nature and emanated from the Home Office not the Ministry of Labour. It was designed to help reunite

victims of the Holocaust with relatives in England. Although laudable in its aims, the DRS was operated in a miserly fashion. Only 5000 Holocaust survivors were ultimately allowed to enter under its remit. For details see London, *Whitehall and the Jews*, pp. 266–9, Kushner, *The Holocaust and the Liberal Imagination*, pp. 229–38 and Cesarani, *Justice Delayed*, pp. 78–88.

269 D. Cesarani et al., *Report on the Entry of Nazi War Criminals and Collaborators into the UK, 1945–1950*, published for the House of Commons, All Party Parliamentary War Crimes Group, November, 1988.
270 Cesarani, *Report on the Entry of Nazi War Criminals*, p. 51.
271 Cesarani, *Report on the Entry of Nazi War Criminals*, p. 23.
272 Cesarani, *Report on the Entry of Nazi War Criminals*, 10/3/46.
273 In fact, there is evidence that the government was, on its own terms, prepared to go to significant lengths to smooth potential racial hostility towards EVWs where it feared it might occur. See Paul, *Whitewashing Britain*, p. 86, Kay and Miles, *Refugees or Migrant Workers?*, p. 172, K. Sword et al., *The Formation of the Polish Community in Great Britain 1939–1950* (London: Caldra House, 1989), pp. 278–89, Tannahill, *European Volunteer Workers in Britain*, pp. 68–70 and Bevan, *The Development of British Immigration Law*, p. 75.
274 Cesarani, *Justice Delayed*, p. 81. This analysis of exclusion has been supported by other experts on the EVW programmes. See Kay and Miles, *Refugees or Migrant Workers?*: 'EVWs were not conceptualised as a single, homogenous category, but were, minimally, dichotomised. Nevertheless, they were all racialised: the discourse of "race" was implicit (as in references to "blood" and "stock") and explicit in the drawing of conclusions about the "suitability" of EVWs, but in combination with evaluations of their social and cultural attributes', p. 123. Also see Paul, *Whitewashing Britain*, pp. 64–89.
275 Cesarani, *Justice Delayed*, p. 73. Also see conclusions of Kay and Miles, *Refugees or Migrant Workers?*, p. 175.
276 Cesarani, *Justice Delayed*, p. 73.

4 Race on the Retreat? The 1950s and 1960s

1 For details of Lysenko's career see V. Soyfer (trans. L. Gruliow and R. Gruliow), *Lysenko and the Tragedy of Soviet Science* (New Brunswick: Rutgers University Press, 1994), D. Joravsky, *The Lysenko Affair* (Cambridge, Massachusetts: Harvard University Press, 1970), Werskey, *The Visible College*, pp. 205–10, McGucken, *Scientists, Society and the State*, pp. 282–4 and Z. Medvedev, *The Rise and Fall of TD Lysenko* (New York: Columbia, 1969). For Lysenko's scientific ideas see T. Lysenko (trans. T. Dobzhansky), *Heredity and its Variability* (New York: Kings Crown Press, 1946).
2 For the interaction between Stalin and Lysenko see K. Rossianov, 'Editing Nature: Joseph Stalin and the "New" Soviet Biology', *ISIS* (1993), 84: 4, 728–45.
3 See Jones, *Science, Politics and the Cold War*, p. 23.
4 See Werskey, *The Visible College*, p. 293.
5 Kohn, *A Reason for Everything*, p. 173.
6 See Barkan, *The Retreat of Scientific Racism*, p. 243.
7 See Huxley MSS, Box 15, Huxley to *Soviet War News*, 2/8/41.

8   Huxley MSS, Huxley to Geoffrey Vevers, 21/12/39 and 28/2/40.
9   *Time Magazine*, 18/7/49.
10 Kohn, *A Reason for Everything*, p. 167. The suspicious death of Vavilov in 1943 would be a key factor in changing Haldane's post-war attitude towards Soviet science.
11 Werskey, *The Visible College*, pp. 206–10.
12 Haldane MSS, 20551, 'The Dark Religions', *The Rationalist Annual*, 1961.
13 Hogben expressed sympathy with Marxism in conversations with Gary Werskey in 1968. He told Werskey: 'My experience in South Africa alone had made me receptive to the Communists as the one political group concerned with four-fifths of the population.' Hogben MSS, File A72, 26/7/68. For Hogben's view on the superior opportunities in Soviet science see Werskey, *Visible College*, p. 205.
14 For details of Hogben's views after this impromptu visit see Huxley MSS, Box 15, Hogben to Huxley, 12/11/40: Hogben argued: 'I've seen Russia, and prefer even Nazi Germany to Stalin's tyranny of squalor, inefficiency and illiteracy.'
15 Werskey, *The Visible College*, p. 295. Jones also argues that Haldane found the Lysenko Affair particularly stressful in *Science, Politics and the Cold War*, p. 27.
16 Haldane MSS, 20594, Draft of 'Lysenko and Genetics', 1940. For analysis of Haldane's perspective see Werskey, *The Visible College*, pp. 208–9 and Kohn, *A Reason for Everything*, pp. 174–87.
17 Haldane MSS, 20548, Paper on Soviet Genetics, 1946.
18 Kohn, *A Reason for Everything*, pp. 182–4 and Werskey, *The Visible College*, p. 293.
19 For a thorough analysis of Haldane's responses to Lysenko see Jones, *Science, Politics and the Cold War*, pp. 27–31.
20 J. Huxley, 'JBS Haldane', *Encounter* (1965), 65: 4, 60.
21 Huxley described the delegations from Western countries as being 'in large part composed of Communists or Communist sympathisers' in J. Huxley, 'Intellect at Wroclaw', *The Spectator*, September 10, 1948, pp. 326–7. Jones briefly addresses the impact of this conference in *Science, Politics and the Cold War*, p. 25.
22 Huxley, *Memories II*, pp. 162–4.
23 Huxley, *Memories II*, p. 162.
24 Huxley, 'Intellect at Wroclaw', p. 326.
25 Huxley, *Memories II*, p. 64.
26 See J. Huxley, 'Is this the End of Science?', *Leader Magazine*, 3/12/49, J. Huxley, 'Soviet Genetics: the Real Issue', *Nature*, 25/6/49, J. Huxley, 'Lysenko Once More', *The Spectator*, 11/11/49, J. Huxley, 'Party Line', *Time Magazine*, 18/7/49, p. 53 and Huxley's monograph on the subject, *Soviet Genetics and World Science: Lysenko and the Meaning of Heredity* (London: Chatto and Windus, 1949). That Huxley wanted to spread the word is evident in his efforts to get this work translated into Swedish, Italian, Dutch, French and Spanish. See Huxley MSS, Box 18, 1–5 July, Huxley to Gustafson, Montalenti, Sirks, Castier and Ocampo.
27 Huxley MSS, Box 17, Huxley to Eryn Hovde, 13/9/48.
28 Huxley MSS, Box 17, Albert Einstein to Huxley, 14/9/48.

29 Haldane MSS, 20535, Haldane to the Provost of UCL, 22/6/48. Jones has argued that scholars outside the USSR often read Lysenkoism 'as purely a response to the rise of racial thinking in Europe', in *Science, Politics and the Cold War*, p. 3.
30 Werskey, *The Visible College*, p. 294.
31 Huxley, *Memories II*, p. 64.
32 Jones, *Science, Politics and the Cold War*, p. 23.
33 Werskey, *The Visible College*, p. 303.
34 Jones, *Science, Politics and the Cold War*, p. 49.
35 M. Banton, *The International Politics of Race* (Cambridge: Polity, 2002), p. 28.
36 Cited in UNESCO, *Statement on Race: an Annotated Elaboration and Exposition of the Four Statements on Race Issued by the UNESCO* (New York: OUP, 1972), p. 1. This publication was the annotated third edition of the UNESCO race statements. The 'annotated elaboration' was written by Ashley Montagu.
37 For details of the handover see Huxley, *Memories II*, p. 67. For Huxley's initial involvement and role, pp. 13–65.
38 UNESCO, *Statement on Race*, pp. 1–6.
39 The committee members were Ernest Beaglehole, Juan Comas, Jan Czechanowski, Franklin Frazier, Morris Ginsberg, Humayun Kabir, Claude Levi-Strauss, Ashley Montagu, L.A. Costa Pinto and Joseph Skold.
40 Huxley, *Memories II*, p. 24. See Jones, *Science, Politics and the Cold War*, p. 68.
41 UNESCO, *Statement on Race*, p. 4.
42 UNESCO, *Statement on Race*, p. x.
43 For a thorough analysis of the first UNESCO Statement on Race see Kohn, *The Race Gallery*, pp. 40–7.
44 'The UNESCO Statement by Experts on Race Problems', 18/7/50, Article Five.
45 UNESCO Statement, Article Four.
46 UNESCO Statement, Article Fourteen.
47 UNESCO Statement, Article Six.
48 UNESCO Statement, Article Nine.
49 UNESCO Statement, Article Ten.
50 UNESCO Statement, Article Nine.
51 UNESCO, *Statement on Race*, p. 105.
52 See Jones, *Science, Politics and the Cold War*, p. 69.
53 Barkan, *The Retreat of Scientific Racism*, p. 342. For a similar analysis see M. Kenny, 'Racial Science in Social Context: John R. Baker on Eugenics, Race, and the Public Role of the Scientist', *ISIS* (2004), 95, 394–419, 405.
54 Kevles, *In the Name of Eugenics*, p. 138.
55 Huxley MSS, Box 18, Huxley to Angell, 17/1/50.
56 Huxley MSS, Box 18, Huxley to Angell, 17/1/50.
57 Huxley MSS, Box 18, Huxley to Angell, 17/1/50. Huxley reiterated these concerns about the absence of biologists and the selection of Ginsberg in a letter to John Maud at UNESCO's Ministry of Education. See Huxley MSS, Box 19, Huxley to Maud, 18/1/50.
58 For analysis see Banton, *The International Politics of Race*, pp. 28–30, Kevles, *In the Name of Eugenics*, p. 138 and Barkan, *The Retreat of Scientific Racism*, pp. 341–3.
59 Huxley MSS, Box 19, Metraux to Huxley, 26/1/51.

60 Banton, *The International Politics of Race*, p. 29. Also see Jones, *Science, Politics and the Cold War*, p. 69.
61 See this letter in *The Times*, 15/8/50. Also see the editorial in *Man*, 'UNESCO on Race' (October 1950), 220, 138.
62 Letter of H.J. Fleure in *Man* (January 1951), 51, 16.
63 Eugenics Society MSS, SA/EUG C106, Fleure to Blacker, 1/8/50.
64 Little in *Man* (January 1951), 51, 17.
65 See for example the comments of Don Hager (Dept of Economics and Social Institutions, Princeton University) in the April 1951 edition regarding the letter of Dr Osman Hill in January 1951.
66 Huxley MSS, Box 19, Huxley to Metraux, 23/2/51.
67 Huxley MSS, Box 19, Metraux to Huxley, 26/1/51. For analysis of the UNESCO Second Statement on Race see Kohn, *The Race Gallery*, pp. 45–7.
68 Huxley MSS, Box 19, Metraux to Huxley, 26/1/51.
69 See Huxley MSS, Box 19, Metraux to Huxley, 26/1/51. Huxley scribbled suggestions on the foot of this letter. They were: Solly Zuckerman, C.D. Darlington, E.B. Ford and R.A. Fisher.
70 Huxley MSS, Box 19, Huxley to Metraux, 23/2/51.
71 Huxley MSS, Box 19, Huxley to Metraux, 23/2/51.
72 *Man* (November 1951), 51, 154.
73 UNESCO, *Statement on Race*. Montagu claimed in his preface to the Second Statement: 'Comparison between the first and second Statements will show that there is, in fact, very little difference between them', p. 138.
74 UNESCO, *Statement on Race*, p. 139.
75 UNESCO, Statement on the Nature of Race and Race Differences by Physical Anthropologists and Geneticists, July 1952 (Second Statement), Article 5. This is being compared to article 9 of the first Statement.
76 UNESCO Second Statement, Article 5 and Article 6.
77 UNESCO Second Statement, Article 5. A belief in the existence of small racial mental differences is also evident in other UNESCO publications in this period. See G. Morant, *The Significance of Racial Differences* (Paris: UNESCO, 1952), pp. 34–43.
78 The issue of Montagu's 'interim statement' was first raised in *Man* in January 1952, p. 9. The Statement from the British delegation was published in *Man* in its May 1952 edition, p. 78. As far as I can see there was actually no difference whatsoever between Montagu's interim statement and the official second Statement of 1952.
79 Montagu responded in a letter to *Man* (April 1952), 52, 63–4.
80 *Man*, Letter from Metraux (March 1952), 52, 47.
81 Bibby was Senior Lecturer at the College of St Mark and St John, University of London, Institute of Education.
82 Cyril Bibby MSS, Cambridge Central Library, File 5/28, Metraux to Bibby, 30/9/53. Bibby was paid $600 to write the study.
83 Bibby MSS, File 5/28, Bibby to Metraux, 6/10/53.
84 Bibby MSS, File 5/28, Metraux to Bibby, 9/10/53.
85 Bibby MSS, File 5/28, Notes.
86 The 'Expert meeting on the promotion of teaching of race questions in primary and secondary schools' was chaired by Alva Myrdal and took place between 19 and 25 September 1955. It considered Bibby's work along with a text aimed at a primary school readership, 'Learning to Live

Together without Hate' by Prof. Charles Hendry. See Bibby MSS, File 5/29 for details of the meetings.
87. Metraux had informed Bibby at the time of the contract offer: 'The manuscript will have to be submitted to National Commissions before publication, with a request for comments, since we are fully aware of the fact that the main difficulty is to fit the text to the particular situations and mentalities of the different countries.' Bibby MSS, File 5/28, Metraux to Bibby, 30/9/53.
88. Bibby MSS, File 5/28, Metraux to Bibby, 6/12/54.
89. Bibby MSS, File 5/28, Metraux to Bibby, 30/9/55.
90. See Huxley MSS, Box 27, Bibby to Huxley, 5/10/58 and Bibby MSS, File 5/28, Metraux to Bibby, 8/1/57.
91. The book finally appeared as C. Bibby, *Race, Prejudice and Education* (London: Heinemann, 1959). There is evidence that UNESCO forced Heinemann to delay the publication ostensibly over issues of permission. See Huxley MSS, Box 27, Bibby to Huxley, 10/10/58.
92. Bibby MSS, File 5/28, Metraux to Bibby, 5/6/57. Of the six changes to the text demanded by UNESCO all (bar one typographical correction) were connected to comments about the USA.
93. Amongst other demands, the Americans called for the removal of Paul Robeson and Josephine Baker from Bibby's list of Negro 'outstanding achievement'. Bibby MSS, File 5/28, Metraux to Bibby, 5/6/57.
94. President's Committee on Civil Rights, Executive Order 9808, 5/12/1946. The Committee criticised the Supreme Court's *Plessy* vs *Ferguson* decision of 1896, which upheld the notion of 'separate but equal'. See S. Truman (ed.), *To Secure These Rights: the Report of Harry S. Truman's Committee on Civil Rights* (Boston and New York: Bedford/St Martins, 2004), pp. 111–13. Call for the elimination of segregation, p. 179.
95. See J.P. Jackson, 'The Scientific Attack on Brown v. Board of Education, 1954–1964', *American Psychologist* (2004), 59: 6, 530–7, 530. Also I. Newby, *Challenge to the Court: Social Scientists and the Defense of Segregation 1954–1966* (Baton Rouge: Louisiana State University Press, 1967), pp. 19–61.
96. See Newby, *Challenge to the Court*, pp. 43–61.
97. Bibby, *Race, Prejudice and Education*, p. 79.
98. Bibby, *Race, Prejudice and Education*, pp. 1, 36–7 and 55.
99. Huxley MSS, Box 27, Bibby to Huxley, 5/10/58.
100. Bibby was particularly loath to accede to the American demand that he remove the name of Paul Robeson from his list of 'eminent Negroes' and instead chose to remove the entire list (Huxley MSS, Box 27, Bibby to Huxley, 5/10/58). Presumably this request was rooted in American hostility towards Robeson's communism and the influence that it may have had on disaffected black youth.
101. Bibby also suspected that the Director General of UNESCO, the Texan Luther H. Evans, was against the book. The election of Evans (who served between 1953 and 1958) may indeed explain Bibby's changing fortunes at UNESCO. See Huxley MSS, Box 27, Bibby to Huxley, 10/10/58.
102. Huxley MSS, Box 27, Bibby to Huxley, 5/10/58 and 10/10/58. Bibby wrote in this letter that he had made 'concession after concession' but had failed to meet American demands.
103. Bibby MSS, File 5/28, Bibby's personal notes.

104 See C. Bibby, *T.H. Huxley: Scientist, Humanist and Educator* (London: Watts, 1959) and C. Bibby, *Scientist Extraordinary: the Life and Scientific Work of Thomas Henry Huxley 1825–95* (New York: St Martins Press, 1972).
105 Huxley MSS, Box 27, Huxley to Bibby, 7/11/58.
106 Huxley MSS, Box 27, Huxley to Bibby, 8/10/58.
107 Bibby, *Race, Prejudice and Education*, p. 64.
108 Huxley MSS, Box 27, Huxley to Bibby, 8/10/58.
109 L. Hogben, 'A Journey Through Ghana', Hogben MSS, A22, 1956.
110 Hogben, *A Journey Through Ghana*, pp. 285–6.
111 Huxley's lecture was published as: J. Huxley, *Man's New Vision of Himself* (Natal: EP and Commercial Printing, 1960).
112 Huxley, *Man's New Vision of Himself*, p. 18.
113 Huxley, *Man's New Vision of Himself*, pp. 17–20.
114 Huxley, *Man's New Vision of Himself*, p. 19.
115 Haldane MSS, 20651, Folio 150, C1955.
116 The 1962 Galton Lecture. Published as J. Huxley, 'Eugenics in Evolutionary Perspective', *Perspectives in Biology and Medicine* (1962), 6: 2, 155–87, 165.
117 Huxley MSS, Box 32, Gates to Huxley, 29/11/61.
118 Hogben MSS, A21, Autobiographical notes, pp. 9–10. Looking back on Hogben's career, Solly Zuckerman remembered him as a determined opponent of apartheid who in the wake of his time in South Africa 'went out of his way to show that he had no racial prejudices', Hogben MSS, A74, Letter from Zuckerman to G.P. Wells (about the career of Hogben), 2/4/76.
119 Huxley, *Man's New Vision of Himself*, p. 17.
120 Gates was sent the second Statement as part of the wide-scale consultation undertaken by UNESCO. There is no record of his comments but he did read and mark the script sent to him. Where the Statement read: 'There is no evidence that race mixture produces disadvantageous results from a biological point of view. The social results of race mixing whether for good or ill, can generally be traced to social factors', Gates indicated his incredulity with a large question mark. Gates MSS, File 11/29, Gates's notes on the UNESCO Second Statement, C1951.
121 Eugenics Society MSS, EUG/C120, Gates to Blacker, 2/11/49.
122 Gates MSS, File 1/12, Keith to Gates, 20/11/50.
123 See Gates MSS. For rejection by the *American Journal of Physical Anthropology*, see Box 1/13, Howells to Gates, 10/6/51. For *Science* rejection see Box 1/13, Editor to Gates, 15/5/51 and for the *Journal of the American Society of Human Genetics* see Box 1/13, Herndon to Gates, 31/3/52.
124 W.W. Howells, the editor of the *American Journal of Physical Anthropology*, explained his rejection of a 1951 article by Gates in this way. Trying to reassure Gates that the reason was not a political one, he outlined that it was his method that was lacking. Describing Gates's paper, Howells noted: 'the whole thing rests on a handful of cases subjectively described and interpreted, and does not represent intensive scientific work'. Gates MSS, Box 1/13, Howells to Gates, 10/6/51. For a detailed analysis of this issue see Schaffer, 'Scientific Racism Again?', pp. 253–78.
125 Gates MSS, Box 1/13, Gates to Howells, 3/4/51.
126 Gates MSS, Box 1/13, Gates to Strandskov, 20/12/51. See Jones, *Science, Politics and the Cold War*, pp. 120–1.

127 Gates MSS, Box 1/12, Brimble to Gates, 15/4/49.
128 Jackson has noted that many scholars in the American South believed that the anti-segregation writings of Gunnar Myrdal were part of a wider communist conspiracy to undermine the United States in Jackson, *Gunnar Myrdal*, p. 292. Similarly, Newby has argued that scientific racists commonly perceived themselves as 'martyr[s] to academic freedom' in Newby, *Challenge to the Court*, p. 65.
129 Gates MSS, Box 1/12, Keith to Gates, 23/2/50.
130 Keith was of course describing the work of T.H. Huxley, not Julian. Gates MSS, Box 1/13, Keith to Gates, 22/5/51.
131 Gates MSS, Box 1/14, Keith to Gates, 12/5/53.
132 Gates MSS, Box 1/14, Keith to Gates, 22/2/53.
133 Gates MSS, Box 1/12, Keith to Gates, 9/5/50.
134 Gates MSS, Box 1/12, Keith to Gates, 27/9/51.
135 Gates MSS, Box 1/14, Hooton to Gates, 18/5/53.
136 Gates MSS, Box 1/12, Meyerhoff to Dr Lorin Mullins (Biological Laboratory, Purdue University), 2/9/49.
137 See Tucker, *The Funding of Scientific Racism*, p. 91.
138 Gates MSS, Box 1/17, Gayre to Gates, 2/7/56. In another letter Gayre wrote: 'on all major points I am entirely of your opinion', Gates MSS, Box 1/17, Gayre to Gates, 4/10/56.
139 Gates MSS, Box 1/17, Gayre to Gates, 4/10/56.
140 Gates MSS, Box 1/18, Gayre to Gates, 4/2/57.
141 Gates MSS, Box 1/19, Gayre to Gates, 27/2/58.
142 Gates MSS, Box 1/19, Gayre to Gates, 18/10/58.
143 Gates MSS, Box 1/19, Gayre to Gates, 8/12/58. Specifically, this outburst related to a conflict between Gates and the Edinburgh University geneticist Sheldon Wolff who had criticised Gates's work in the *Scotsman*. In another letter Gayre described Wolff as 'a common little Cockney Jew'. Gates MSS, Box 1/19, Gayre to Gates, 27/2/58.
144 Gates MSS, Box 1/19, Gayre to Gates, 29/7/58.
145 See Jones, *Race, Science and Politics*, p. 122.
146 Newby, *Challenge to the Court*, pp. 185–93 and Jackson, *Gunnar Myrdal*, p. 252.
147 Newby, *Challenge to the Court*, p. 186.
148 Newby, *Challenge to the Court*, pp. 191–212.
149 Gates MSS, Box 1/16, W.J. Simmons to Gates, 24/5/55.
150 In 1958, the pro-segregation businessman Edward Benjamin financed Gates's research field trip to Australia. See Gates MSS, Box 1/19, Benjamin to Gates, 14/2/58. Gates was happy enough to voice an opinion on the segregation issue, only he wanted to emphasise his own lack of ideological partiality as he did so. Presumably, he considered that his lack of political affiliation added credence to his campaigning. Keeping his cards close to his chest, Gates was maybe also mindful that direct ties with pro-segregation organisations would put an end once and for all to his chances of securing employment in most American universities. A letter from the Chief of the Association of Citizen Councils of Mississippi highlights their clandestine relationship with Gates. It told him: 'You cannot imagine how much it means to us to have the sympathetic understanding of men like yourself...You are correct, I think, in feeling that you can be more

helpful by not becoming a member' (Gates MSS, Box 1/17, Simmons to Gates, 28/6/56).

151 For analysis of the *Mankind Quarterly's* establishment see Jones, *Science, Politics and the Cold War*, pp. 123–5.

152 For a thorough analysis of Garrett's racial views see Tucker, *The Funding of Scientific Racism*, pp. 79–81. Tucker describes Garrett as 'arguably the most eminent of the scientific segregationists', p. 79. Newby describes the role of Garrett in the 1951 *Davis* v. *County School Board of Prince Edward County* case in *Challenge to the Court*, pp. 53–4.

153 See Newby, *Challenge to the Court*, p. 321.

154 Gayre's editorial notes, the *Mankind Quarterly* (1960), 1: 1, 4 and the *Mankind Quarterly* (1960), 1: 2, 108.

155 See Gates, *The Emergence of Racial Genetics*.

156 Gates MSS, Box 1/21, Gayre to Gates, 18/5/60.

157 Gates MSS, Box 1/21, Gayre to Gates, 15/4/60.

158 This donation is described in a letter from Garrett to Gates. Gates MSS, Box 1/21, 2/4/60. Gayre wrote in the same period expressing hopes that Draper would fund the entire costs of the *Quarterly*'s first year. Gates MSS, Box 1/21, Gayre to Gates, 7/4/60.

159 See D. Blackmon, 'Silent Partner: How the South's Fight to Uphold Segregation Was Funded Up North', *Wall Street Journal*, 6 Nov. 1999 and Tucker, *The Funding of Scientific Racism*. Also see Y. Katagiri, *The Mississippi State Sovereignty Commission: Civil Rights and States' Rights* (Jackson: University Press of Mississippi, 2001). Draper funded numerous scientific projects where he perceived that they may prove the validity of segregationist policy. The major beneficiary from Draper's will was the 'Pioneer Fund', a pro-eugenics organisation co-founded by Draper himself in 1937, which pursued a range of pro-segregation and eugenic policies. For analysis of the fund see M. Kenny, 'Toward a Racial Abyss: Eugenics, Wickliffe Draper, and the Origins of the Pioneer Fund', *Journal of the History of the Behavioural Sciences* (2002), 38: 3, 259–83.

160 For the history of the IAAEE see Tucker, *The Funding of Scientific Racism*, pp. 78–101 and Newby, *Challenge to the Court*, p. 119.

161 See Tucker, *The Funding of Scientific Racism*, p. 90. Tucker notes here that IAAEE members were at one stage entitled to a free copy of the *Mankind Quarterly*.

162 The journal was seemingly not economically viable. Gayre informed Gates that at the journal's beginning it had only secured 40 subscriptions. Gates MSS, Box 1/21, Gayre to Gates, 27/4/60.

163 Comas contributed to both of the first two UNESCO statements on race and wrote a pamphlet on the subject for the organisation. See J. Comas, *Racial Myths* (Paris: UNESCO, 1951). The article under discussion here was J. Comas, ' "Scientific" Racism Again?', *Current Anthropology* (1961), 2: 4, 303–40.

164 Gayre's editorial appeared as: 'The Mankind Quarterly under Attack', the *Mankind Quarterly* (1961), 2: 2, 42–7.

165 See Newby, *Challenge to the Court*, p. 320.

166	D. Purves, 'The Evolutionary Basis of Race Consciousness', the *Mankind Quarterly* (1960), 1: 1, 51–4.
167	For example, see R. Hall, 'Zoological Subspecies of Man', the *Mankind Quarterly* (1960), 1: 2, 118–19, S.D. Porteus, 'Ethnic Group Differences', the *Mankind Quarterly* (1961), 1: 3, 187–200, and H. Garrett, 'The Equalitarian Dogma', the *Mankind Quarterly* (1961), 1: 4, 253–7. For the *Quarterly*'s attitude to South Africa see R. Gayre, 'The Bantu Homelands of the Northern Transvaal', the *Mankind Quarterly* (1962), 3: 2, 98–112.
168	In particular, two articles considering 'The Jewish Role in the American Elite' kept an anti-Semitic agenda present in the journal. See N. Weyl, 'The Jewish Role in the American Elite', the *Mankind Quarterly* (1962), 3: 1, 26–36 and N. Weyl, 'The Ethnic and National Characteristics of the US Elite', the *Mankind Quarterly* (1961), 1: 4, 242–52.
169	For the anti-Semitism of Garrett see Tucker, *The Funding of Scientific Racism*, p. 80.
170	In 1978 the *Mankind Quarterly* finally moved to its more logical home in the US, when the white supremacist Roger Pearson assumed the editorship. See Kohn, *The Race Gallery*, pp. 53–4 and Tucker, *The Funding of Scientific Racism*, pp. 88–90.
171	Gayre had previously published his heraldry journal, *the Armorial*, from Edinburgh.
172	Gates MSS, Box 1/21, Gayre to Gates, 24/3/60.
173	Gates MSS, Box 1/21, Gayre to Gates, 23/5/60. The *Mankind Quarterly* did finally attract Sir Charles Darwin, a non-affiliated anthropologist from Cambridge, onto the editorial board. Darwin had been President of the Eugenics Society between 1953 and 1959. Jones has noted that the geneticist C.D. Darlington also became involved with the journal in *Science, Politics and the Cold War*, p. 128.
174	Gates MSS, Box 1/21, Gayre to Gates, 15/4/60.
175	Haldane MSS, 20522, Folios 191–194, 'A Text for American Fascists', 1963.
176	Haldane did accuse the authors of being 'ignorant of much of the relevant literature' and of making 'contradictory statements'.
177	Haldane MSS, 20561, Folio 213, 'Notes on Race', C1955.
178	Haldane MSS, 20604, 'Essay on Race', C1945–6.
179	Haldane MSS, 20561, Folio 212, 'Notes on Race', C1955.
180	Analyses which address this immigration are too prevalent to cite. Major works include P. Fryer, *Staying Power: the History of Black People in Britain* (London: Pluto, 1984), E. Rose et al., *Colour and Citizenship: a Report on British Race Relations* (London: Oxford University Press, 1969), Miles, *Racism and Migrant Labour*, T. and M. Phillips, *Windrush: the Irresistible Rise of Multi-Racial Britain* (London: HarperCollins, 1998), Paul, *Whitewashing Britain*, C. Peach, *West Indian Migration to Britain: a Social Geography* (London: Oxford University Press, 1968), I. Spencer, *British Immigration Policy since 1939* (London: Routledge, 1997) and R. Hansen, *Citizenship and Immigration in Post War Britain: the Institutional Origins of a Multi-Cultural Nation* (Oxford: Oxford University Press, 2000).
181	The Act was perceived as a necessary confirmation of the unity of British citizenship throughout the Commonwealth after Indian independence and the Canadian Nationality Act of 1946 had created citizenship separate from

	Britain. For analysis of this legislation see Paul, *Whitewashing Britain*, p. 16, Spencer, *British Immigration Policy since 1939*, p. 55 and Bevan, *The Development of British Immigration Law*, p. 76.
182	Spencer, *British Immigration Policy*, p. 55 and Hampshire, *Citizenship and Belonging*, p. 47.
183	PRO, PREM11/3238, Norman Brook to Macmillan, 9/10/61. Some analysts have dismissed the importance of the Commonwealth within the thinking of politicians in this period (notably John Solomos and Ambalavaner Sivanandan). Hansen, however, has challenged this view, arguing that there is evidence to suggest that many within the British government were genuinely concerned about the implications of restrictionist legislation on Commonwealth relations in *Citizenship and Immigration*, p. 60. It is unsurprising that the most vehement internal opposition to restriction came from the Colonial and Commonwealth Relations Offices. Some historians have argued that it was objections within these ministries that prevented the early enactment of immigration legislation in the 1950s. Deakin has concluded: 'The private pressure of the Commonwealth Relations Office and Colonial Office seems to have outweighed the arguments advanced by the Home Office and the joint working party', in Rose *et al.*, *Colour and Citizenship*, p. 210. A 1960 memorandum from a civil servant within the Colonial Office highlighted the fears of this Ministry about even hidden racial restriction. 'However disguised, such restrictions would be interpreted as anti-colour measures and this would weaken the Commonwealth and damage the reputation of the UK internationally', PRO, CO1031/3932, Colonial Office Memorandum (Undated), 1960.
184	See Hansen, *Citizenship and Immigration*, pp. 13–35.
185	Hampshire, *Citizenship and Belonging*, p. 33.
186	Stone, 'Race in British Eugenics', pp. 415–16.
187	G. Bertram, 'West Indian Immigration', *Eugenics Society Broadsheet*, 1 (1958), 6. Charles Darwin made a similar argument in the Broadsheet's preface, noting: 'In too many subjects present policy is directed by a benevolent, often a merely sentimental, wish to benefit everybody, without giving any thought to the inherent qualities of humanity which may well defeat the intended aims', p. 3.
188	Bertram, 'West Indian Immigration', pp. 7 and 20.
189	Bertram argued in the pamphlet that it 'would be remarkable' if it was shown that all races shared the same 'basic mental potentiality', p. 17.
190	Bertam, 'West Indian Immigration', p. 22.
191	Bertam, 'West Indian Immigration', p. 16.
192	Bertram, 'West Indian Immigration', p. 18.
193	Bertram, 'West Indian Immigration', p. 20.
194	Eugenics Society MSS, D104, Weyher to Bertram, 4/3/61.
195	Eugenics Society MSS, D104, Weyher to Bertram, 11/4/61.
196	Gates recalled his meetings with Draper in correspondence with Arthur Keith. See Keith MSS, KL1, Gates to Keith, 1/12/54.
197	Eugenics Society MSS, D104, Weyher to Bertram, 11/4/61.
198	Eugenics Society MSS, D104, Gates to Bertram, 16/3/61.
199	Eugenics Society MSS, D104, Bertram to Weyher, 7/6/61.
200	Eugenics Society MSS, D104, Weyher to Bertram, 6/7/61.

201 Eugenics Society MSS, D104. The Liverpool project was titled: 'A Study of the Genetics of Quantitatively Varying Human Interpopulation Differences'. The Birmingham project was titled: 'Inquiry into the Fertility of Immigrants'. Both projects made provisional reports to Bertram for Draper in 1963.
202 Keith MSS, KL1, Gates to Keith, 1/12/54.
203 For the rise of sociology in British universities in this period see A. Halsey, *A History of Sociology in Britain: Science, Literature and Society* (Oxford: OUP, 2004), pp. 70–112.
204 See R. Miles, *Racism after 'Race' Relations* (London and New York: Routledge, 1993), p. 111.
205 See Little, *Negroes in Britain*, R. Glass, *Newcomers: the West Indians in London* (London: George Allen and Unwin, 1960), S. Patterson, *Dark Strangers: A Study of West Indians in London* (London: Penguin, 1965), Richmond, *Colour Prejudice in Britain*, A. Richmond, *The Colour Problem* (Harmondsworth: Penguin, 1955), M. Banton, *The Coloured Quarter: Negro Immigrants in an English City* (London: Jonathan Cape, 1955) and M. Banton, *White and Coloured: the Behaviour of British People towards Coloured Immigrants* (London: Jonathan Cape, 1959).
206 Despite noting the influence of American sociologists on their British colleagues, Rich has argued that the influence of the American model should not be overemphasised. See Rich, *Race and Empire*, pp. 191–2.
207 See King, *Race Culture*, pp. 21–48, V. Williams, *The Social Sciences and Theories of Race* (Urbana and Chicago: University of Illinois Press, 2006), pp. 120–6, Jackson, *Gunnar Myrdal*, pp. 272–311 and J. Jackson, 'The Scientific Attack on Brown v. Board of Education, 1954–1964', *American Psychologist* (2004), 59: 6, 530–7, 530.
208 Little, *Negroes in Britain*. Anthony Richmond described Little's research as 'the pioneering study' in 'Sociological and Psychological Explanations of Racial Prejudice: Some Light on the Controversy from Recent Researches in Britain', *Pacific Sociological Review* (1961), 4: 2, 63–8, 63. Rich has cited the importance of Little spending a sabbatical year in Fisk University in 1949 in *Race and Empire*, p. 191.
209 Banton studied for his PhD under Little's supervision, looking at the black population of Bristol. Patterson carried out her study of Brixton's black population under Little between 1955 and 1958. For analysis of the Edinburgh school see Rich, *Race and Empire*, pp. 191–3.
210 Richmond, *Colour Prejudice in Britain*.
211 For the influence of American social science on Richmond and Little see M. Clapson, 'The American Contribution to the Urban Sociology of Race Relations in Britain from the 1940s to the early 1970s', *Urban History* (2006), 33: 2, 253–73, 256–9. Also see Rose *et al.*, *Colour and Citizenship* which claimed that it had taken its inspiration from Gunnar Myrdal's US study, pp. 1–2.
212 Little, *Negroes in Britain*, p. xi.
213 K. Little, 'Behind the Colour Bar', *Current Affairs* (1950), 118, 1–19, 17.
214 Richmond, 'Colour Prejudice in Britain', p. 15. Richmond described racial prejudice as 'an attitudinal phenomenon' in A. Richmond, 'Sociological and Psychological Explanations of Racial Prejudice', p. 67.
215 Glass, *Newcomers*, p. xi.

216 P. Mason, *Race Relations: a Field of Study Comes of Age* (London: Lucas and Co, 1968), p. 5.
217 Rich has noted Kenneth Little's attempts to persuade the Labour Party to legislate against the Colour Bar in Britain. See Rich, *Race and Empire*, p. 191.
218 Glass, *Newcomers*, p. 214.
219 N. MacKenzie (ed.), C. Senior and D. Manley, *The West Indian in Britain* (London: Fabian Colonial Bureau, 1956), p. 7. For analysis of the Senior and Manley report see Rich, *Race and Empire*, pp. 180–1.
220 Glass, *Newcomers*, p. 229.
221 See Carter, *Realism and Racism*, p. 65.
222 For example, see the work of the *Institute of Race Relations* under the leadership of Philip Mason. This organisation tried to harness the views of various academic communities and push for anti-discrimination legislation. For analysis see Rich, *Race and Empire*, pp. 196–200.
223 Banton, *The Coloured Quarter*, p. 244.
224 Banton, *The Coloured Quarter*, p. 249.
225 Patterson, *Dark Strangers*, p. 345.
226 Foot, *Immigration and Race*, p. 133.
227 M. Banton, *Race Relations* (London: Tavistock, 1967), p. 385.
228 Glass, *Newcomers*, p. 237.
229 Even as they mostly believed that differences were due to social and not biological factors, Britain's leading sociologists still often articulated the idea that black immigrants were significantly different from other Britons in terms of their attitudes and demeanour. Richmond argued, for example, that black immigrants had very different sexual values in *Colour Prejudice in Britain*, pp. 79–80. Banton, in *The Coloured Quarter*, credited black immigrants with a 'cheerful and spontaneous temperament', p. 169.
230 Hampshire, *Citizenship and Belonging*, p. 126.
231 Hampshire, *Citizenship and Belonging*, p. 126.
232 Glass, *Newcomers*, pp. 123–4. For further analysis see Hampshire, *Citizenship and Belonging*, pp. 111–49.
233 Senior and Manley, *The West Indian in Britain*, p. 16.
234 Liverpool Central Archive, British Council of Churches MSS, Informal Conference on Coloured Workers, Liverpool, 2–4 April, 1951, File H325.
235 Hampshire, *Citizenship and Belonging*, p. 141.
236 PRO, CAB129/81, Report by Committee of Ministers, 22/6/56. See Paul, *Whitewashing Britain*, pp. 123–9 and Kay and Miles, *Refugees or Migrant Workers?*, pp. 122–4 for the importance placed on the perceived racial stock of potential immigrants.
237 PRO, CO1032/195, Inter-Departmental Working Party Report on the Immigration of Indians and Pakistanis, 17/4/57.
238 Some analysts have argued that anti-black discourse in the USA and South Africa contributed to allegations of black sexual criminality in Britain. See Phillips and Phillips, *Windrush*, pp. 163–4.
239 Glass, *Newcomers*, p. 150. Banton also wrote about the prevalence of this belief noting that he had found no evidence from his research to support such claims. See Banton, *The Coloured Quarter*, p. 161.

240 See E. Pilkington, *Beyond the Mother Country: West Indians and the Notting Hill White Riots* (London: Tauris, 1988), pp. 92–4. Also see Hampshire, *Citizenship and Belonging*, pp. 135–8.
241 Boyle MSS, Brotherton Archive, University of Leeds, 660/14401, Cress to Boyle, 5/9/58.
242 PRO, CO1028/22, Draft Report of the Working Party on Coloured People Seeking Employment in the UK, 28/10/53.
243 PRO, CO1032/195, Inter-Departmental Working Party Report on the Immigration of Indians and Pakistanis, 17/4/57.
244 PRO, CO1032/195, Report prepared for government response to adjournment motion by H. Hynd, 3/4/58.
245 See Hoch, *White Hero Black Beast*, pp. 43–64 and Gilroy, *Between Camps*, pp. 196–7 for analysis of perceptions of black super-masculinity.
246 Foot, *Immigration and Race*, p. 239.
247 Richmond, *Colour Prejudice in Britain*, p. 79.
248 Richmond, *Colour Prejudice in Britain*, p. 80.
249 Patterson, *Dark Strangers*, p. 338.
250 Glass, *Newcomers*, p. 86.
251 Richmond, 'Economic Insecurity and Stereotypes as Factors in Colour Prejudice', p. 154.
252 PRO, PREM11/824, Cabinet Paper by Eden, 28/10/55.
253 PRO, AST7/1614, Report from the National Assistance Board, 21/5/58.
254 PRO, PREM11/824, Cabinet Minutes, 2/11/55.
255 PRO, CAB129/81, Report by Committee of Ministers, 22/6/56.
256 Paul has argued: 'Skin colour was considered to be an unmodifiable racial characteristic', in Paul, *Whitewashing Britain*, p. 129. Also see Carter, *Realism and Racism*, p. 117 and H. Goulbourne, *Ethnicity and Nationalism in Post Imperial Britain* (Cambridge: Cambridge University Press, 1991), pp. 122–5.
257 A. Sivanandan, *A Different Hunger: Writings on Black Resistance* (London: Pluto Press, 1982), p. 108.
258 Glass, *Newcomers*, p. 119.
259 See Paul, *Whitewashing Britain*, pp. 90–110, Goulbourne, *Ethnicity and Nationalism*, p. 115, Holmes, *John Bull's Island*, pp. 251–4, Miles, *Racism and Migrant Labour*, pp. 121–50 and J. Solomos, *Race and Racism in Britain* (Basingstoke: Macmillan, 1989), pp. 40–3.
260 A. Dummett and M. Dummett, 'The Role of Government in Britain's Racial Crisis' in C. Husband (ed.), *Race in Britain: Continuity and Change* (London: Hutchison, 1982), p. 102.
261 PRO, CAB129/102, Note by G. Lloyd George, 18/8/55.
262 See Paul, *Whitewashing Britain*, L. Curtis, *Nothing but the Same Old Story*, in R. Swift and S. Gilley (eds), *The Irish in the Victorian City* (London: Croom Helm, 1985) and M. Hickman, *Religion, Class and Identity: the State, the Catholic Church and the Education of the Irish in Britain* (Aldershot: Avebury, 1995), pp. 3–16.
263 This point has been emphasised by Paul: '...though perceived as a distinct community of Britishness, inferior to the domestic, Irish migrants apparently ranked higher on the imperial scale than British subjects of colour'. See Paul, *Whitewashing Britain*, p. 107.

264 For details of this agreement see Paul, *Whitewashing Britain*, p. 109.
265 PRO, DO175/121, Report by D.M. Cleary on a memorandum by the Home Secretary, 28/9/61.
266 PRO, CO1028/22, Report of the Working Party on Coloured People Seeking Employment in the UK, 4/12/53.
267 See M. Freeman and S. Spencer, 'Immigration Control, Black Workers and the Economy', *British Journal of Law and Society* (1979), 6: 1, 53–81.
268 PRO, CO1028/22, Colonial Office Memorandum, 24/3/53.
269 PRO, CO1032/195, Watt minute, 8/9/58.
270 PRO, CO1032/195, Memorandum from Secretary of State for the Colonies to all WIO governments. September, 1958.
271 See Spencer, *British Immigration Policy since 1939*, p. 50. Paul has argued that the Indian and Pakistani governments were 'extremely cooperative', in Paul, *Whitewashing Britain*, p. 152.
272 PRO, CO1032/304, Birmingham address by Manley, 11/6/61.
273 Deakin, in Rose *et al.*, *Colour and Citizenship*, p. 218. See Spencer, *British Immigration Policy since 1939*: '…it would have been politically suicide and economically costly for the Jamaican, or any Caribbean, government to attempt to interfere with the flow of migration', p. 103.
274 See Solomos, *Race and Racism in Britain*, p. 61 and Carter, *Realism and Racism*, p. 122. For the wider legal significance of the Act see Hampshire, *Citizenship and Belonging*, p. 12.
275 Dummett, and Dummett, 'The Role…', p. 103. Similar views have been expressed by a disparate range of scholars of post-war British racism. See Fryer, *Staying Power*, pp. 372–99, Hansen, *Citizenship and Immigration*, pp. 11–114, Z. Layton Henry, *The Politics of Race in Britain* (London: Social Science Research Board, 1980), pp. 39–43, Paul, *Whitewashing Britain*, pp. 111–30, Spencer, *British Immigration Policy since 1939*, pp. 129–33, Solomos, *Race and Racism in Britain*, pp. 61–3, Foot, *Immigration and Race in British Politics*, pp. 133–64, Deakin, in Rose *et al.*, *Colour and Citizenship*, pp. 553–78, Holmes, *John Bull's Island*, pp. 262–9 and Goulbourne, *Ethnicity and Nationalism*, pp. 87–125.
276 See Butler's defence of the Bill in Parliament denying any racist intent: 'It affects Canadians, Australians and New Zealanders in the same way as it affects Indians, Pakistanis and West Indians', *Hansard*, Vol: 654, Col: 1192, 27/2/62.
277 See Holmes, *A Tolerant Country?*, pp. 55–6 and Spencer, *British Immigration Policy since 1939*, p. 127.
278 Dummett and Dummett, 'The Role…', p. 102.
279 PRO, CAB21/4774, Memorandum for the Home Secretary, October 1961.
280 PRO, CO1032/304, Notes by the Working Party on Immigration, 28/4/61. The term 'Old Commonwealth countries' described Australia, New Zealand, Canada and South Africa, countries from where any immigration could largely be expected to be white.
281 PRO, CAB21/4774, Report by the Working Party on Immigration, 28/7/61.
282 The influence of the social scientists can also be seen in the government's decision to enact legislation to prevent discrimination in the Race Relations Acts of 1965 and 1968. Rich has highlighted the role of the Institute of Race Relations in shaping this legislation in Rich, *Race and Empire*, p. 200.

5 Epilogue

1. Kohn, *The Race Gallery*, pp. 60–2. The text under discussion was published as J. Baker, *Race* (London: Oxford University Press, 1974).
2. C.P. Blacker MSS, Box 2, Baker to Blacker, 17/4/51. For a thorough analysis of Baker's motivations see Kenny, 'Racial Science', pp. 409–11. For Baker's career see Kenny, 'Racial Science', pp. 395–7 and E. Wilmer and P. Brunet, 'John Randal Baker, 1900–84', *Biographical Memoirs of the Fellows of the Royal Society* (1985), 31, 33–63.
3. See Kenny, 'Racial Science', pp. 412–14.
4. Blacker MSS, Box 2, Baker to Blacker, 1/5/69.
5. Blacker MSS, Box 2, Baker to Blacker, 10/5/69.
6. Blacker MSS, Box 2, Baker to Blacker, 1/5/69. Medawar's comments ultimately appeared on the sleeve of the book itself. For analysis of Medawar's intervention see Kohn, *The Race Gallery*, p. 61. Also see Kenny, 'Racial Science', p. 413.
7. See Kohn, *The Race Gallery*, pp. 60–2.
8. Baker, *Race*, p. 118. For Baker's explanation of 'taxa', pp. 118–47.
9. Baker, *Race*, p. 427.
10. Baker, *Race*. On the productive environment of Northern Europe, p. 428. On the relative primitiveness of Aborigines, p. 430.
11. Baker, *Race*, p. 59.
12. Baker, *Race*, p. 61.
13. Blacker MSS, Box 2, Blacker to Baker, 1/5/69.
14. Kenny has recorded Baker's support for white settlers in South Africa and Rhodesia and his opposition to black immigration to Britain in 'Racial Science', p. 410.
15. Blacker MSS, Box 2, Baker to Blacker, 10/5/69.
16. Blacker MSS, Box 2, Baker to Blacker, 17/7/73.
17. Blacker MSS, Box 2, Baker to Blacker, 17/8/69.
18. Hampshire, *Citizenship and Belonging*, pp. 179–98.

Select Bibliography

a) Primary sources

1) Archives:

i) Public Record Office, Kew, London

Home Office	HO
Dominions Office	DOM
Cabinet Office	CAB
Prime Minister's Office	PREM
Dept of Employment	AST
Colonial Office	CO
Foreign Office	FO
Ministry for War	WO

ii) Private collections

Wellcome Archive, Wellcome Institute, London
Eugenics Society MSS
F. Parkes Weber MSS
Charles Singer MSS
C.P. Blacker MSS

London Metropolitan Archive
Board of Deputies of British Jews MSS

Special Collections, University of Birmingham
Lancelot Hogben MSS

London School of Economics Archive
Charles Seligman MSS

Hartley Archive, University of Southampton
Charles Singer MSS
Mount Temple MSS

Special Collections, University Central Library, Cambridge
Redcliffe Salaman MSS
Cyril Bibby MSS
Alfred Haddon MSS

Liddell Hart Archive, Kings College London
Reginald Ruggles Gates MSS

Liverpool Central Archive
The Fletcher Report, 1930: 'Report on an Investigation into the Colour Problem in Liverpool and Other Ports' for the Liverpool Association for the Welfare of Half-Caste Children
British Council of Churches MSS

Liverpool University Archive
Eleanor Rathbone MSS

Woodson Research Centre, Rice University, Houston
Julian Huxley MSS

Scottish National Archive
J.B.S. Haldane MSS

Royal College of Surgeons Archive
Arthur Keith MSS

University College London Archive
Karl Pearson MSS

2) Printed material

i) Parliamentary records

Hansard:
House of Commons Parliamentary Debates (1902–1962)

Statutes:
Aliens Act (1905)
Aliens Restriction Act (1914)
Aliens Restriction Amendment Act (1919)
British Nationality Act (1948)
Commonwealth Immigrants Act (1962)

Orders:
Aliens Order (1920)
Special Restriction (Coloured Alien Seamen) Order (1925)

Reports:
All Party Parliamentary War Crimes Group: Report on the Entry of Nazi War Criminals and Collaborators into the UK, 1945–50 (1988)
The Report of the Royal Commission on Population (1949)

Reports of the Biological and Medical Committee, Papers of the Royal Commission on Population, Vol: 4 (1950)

ii) Newspapers and journals
The Times (1950)
Man (1951–2)
The Keys: the Journal of the League of Coloured Peoples (1935)
San Francisco Examiner (1938)
Jewish Chronicle (1919)
Daily Mail (1958)
The Listener (1960)
Picture Post (1940)
Time Magazine (1949)
The Tribune (1940)
London Calling (1943)
Leader Magazine (1949)
Mankind Quarterly (1960–5)

3) Books and articles

Aikman, K., 'Race Mixture', *Eugenics Review* (1933), 25: 3, 161–6.

Alderman, G., *The Jewish Community and British Politics* (Oxford: Clarendon, 1983).

Alexander, C., 'Beyond Black: Re-Thinking the Colour/Culture Divide', *Ethnic and Racial Studies* (2002), 25: 4, 552–71.

Andrade, E. and Huxley, J., *Simple Science* (Oxford: Basil Blackwood, 1934).

Armstrong, C., *The Survival of the Unfittest* (London: C.W. Daniel, 1927).

Baker, J., *Race* (London: Oxford University Press, 1974).

Baker, J., *Julian Huxley: Scientist and World Citizen 1887–1975* (Paris: UNESCO, 1978).

Banton, M., *The Coloured Quarter: Negro Immigrants in an English City* (London: Jonathan Cape, 1955).

Banton, M., *White and Coloured: the Behaviour of British People Towards Coloured Immigrants* (London: Jonathan Cape, 1959).

Banton, M., *Race Relations* (London: Tavistock, 1967).

Banton, M., *Racial Theories* (Cambridge: Cambridge University Press, 1998) (first edition 1987).

Banton, M., 'Progress in Ethnic and Racial Studies', *Ethnic and Racial Studies* (2001), 24: 2, 173–94.

Banton, M., *The International Politics of Race* (Cambridge: Polity, 2002).

Barkan, E., *The Retreat of Scientific Racism: Changing Concepts of Race in Britain and the United States Between the World Wars* (Cambridge: Cambridge University Press, 1992).

Beddoe, J., *The Races of Britain: a Contribution to the Anthropology of Western Europe* (London: Trubner and Co., 1885).

Benedict, R., *Race, Science and Politics* (New York: New Age Books, 1940).

Bentwich, N., *The Rescue and Achievement of Refugee Scholars: the Story of Displaced Scholars and Scientists 1933–52* (The Hague: Nijhoff, 1953).

Bertram, G., 'West Indian Immigration', *Eugenics Society Broadsheet*, 1 (1958).

Bevan, V., *The Development of British Immigration Law* (London: Croom Helm, 1986).
Bibby, C., *Race, Prejudice and Education* (London: Heinemann, 1959).
Bibby, C., *T.H. Huxley: Scientist, Humanist and Educator* (London: Watts, 1959).
Bibby, C., *Scientist Extraordinary: the Life and Scientific Work of Thomas Henry Huxley 1825–95* (New York: St Martins Press, 1972).
Blacker, C.P., 'Galton's Views on Race', *Eugenics Review* (1951), 43: 1, 22.
Blacker, C.P., '"Eugenic" Experiments Conducted by the Nazis on Human Subjects', *Eugenics Review* (1952), 44: 1, 9–19.
Caradog Jones, D., *The Economic Status of Coloured Families in the Port of Liverpool* (Birkenhead: Woolman and Sons, 1940).
Carter, B., *Realism and Racism: Concepts of Race in Sociological Research* (London and New York: Routledge, 2000).
Cesarani, D., 'Anti-Alienism in England after the First World War', *Immigrants and Minorities* (1987), 6: 1, 5–29.
Cesarani, D., 'An Embattled Minority: the Jews in Britain during the First World War', *Immigrants and Minorities* (1989), 8: 1–2, 61–81.
Cesarani, D., *Justice Delayed: How Britain Became a Refuge for Nazi War Criminals* (London: Heinemann, 1992).
Cesarani, D. and Kushner, T., *The Internment of Aliens in Twentieth Century Britain* (London: Frank Cass, 1993), pp. 25–52.
Cheyette, B., *Constructions of 'the Jew' in English Literature and Society: Racial Representations 1875–1945* (Cambridge: Cambridge University Press, 1993).
Clapson, M., 'The American Contribution to the Urban Sociology of Race Relations in Britain from the 1940s to the early 1970s', *Urban History* (2006), 33: 2, 253–73.
Clark, R., *The Rise of the Boffins* (London: Phoenix, 1962).
Clark, R., *The Huxleys* (London: Cox and Wyman, 1968).
Clark, R., *The Life and Work of JBS Haldane* (London: Hodder and Stoughton, 1968).
Cockett, R., *Twilight of Truth: Chamberlain, Appeasement and the Manipulation of the Press* (London: Weidenfeld and Nicholson, 1989).
Comas, J., *Racial Myths* (Paris: UNESCO, 1951).
Comas, J., '"Scientific" Racism Again?', *Current Anthropology* (1961), 2: 4, 303–40.
Cooper, R., *Refugee Scholars: Conversations with Tess Simpson* (Leeds: Moorland, 1992).
Cooper, R., *Retrospective Sympathetic Affection: a Tribute to the Academic Community* (Leeds: Moorland, 1996).
Crew, F., *Organic Inheritance in Man* (London: Oliver and Boyd, 1927).
Crowson, N., 'The British Conservative Party and the Jews during the late 1930s', *Patterns of Prejudice* (1995), 29: 2, 15–32.
Crowson, N., *Facing Fascism: the Conservative Party and the European Dictators 1935–40* (London: Routledge, 1997).
Crowson, N. (ed.), *Fleet Street, Press Barons and Politics: the Journals of Collin Brooks 1932–40* (Cambridge: Cambridge University Press, 1998).
Crowther, J., *The Social Relations of Science* (London: Macmillan, 1941).
Crowther, J., *Science at War* (London: DSIR, 1947).
Darnell, R., *And Along Came Boas: Continuity and Revolution in American Anthropology* (Amsterdam: J. Benjamins, 1998).

Davenport, C. and Steggarda, M. et al., *Race Crossing in Jamaica* (Washington: Carnegie Institute, 1929).
Davies, M. and Hughes, A.G., *An Investigation into the Comparative Intelligence and Attainments of Jewish and Non-Jewish School Children* (Cambridge: Cambridge University Press, 1927).
Doane, M., *Femmes Fatales: Feminism, Film, Psychoanalysis* (London: Routledge, 1991).
Dover, C., 'We Europeans', *Nature* (1935), 136: 3445, 736–7.
Dubow, S., *Scientific Racism in Modern South Africa* (Cambridge and New York: Cambridge University Press, 1995).
Dummett, A. and Dummett, M., 'The Role of Government in Britain's Racial Crisis' in Husband, C. (ed.), *Race in Britain: Continuity and Change* (London: Hutchison, 1982).
Ernst, W. and Harris, B. (eds), *Race, Science and Medicine, 1700–1960* (London: Routledge, 1999).
Elliot Smith, G., *Human History* (London: Cape, 1930).
Elliot Smith, G., *The Diffusion of Culture* (London: Watts and Co, 1933).
Endelman, T., *The Jews of Britain 1656–2000* (Berkeley and London: University of California Press, 2002).
Endelman, T., 'Anglo-Jewish Scientists and the Science of Race', *Jewish Social Studies* (2004), 11: 1, 52–92.
Evans, N., 'The South Wales Race Riots of 1919', *Llafer* (1983), 3, 5–29.
Evans-Gordon, W., *The Alien Immigrant* (London: William Heinemann, 1903).
Feldman, D. (ed.), *Englishmen and Jews: Social Relations and Political Culture 1840–1914* (New Haven and London: Yale University Press, 1994).
Foot, P., *Immigration and Race in British Politics* (London: Penguin, 1965).
Foucault, M., *The History of Sexuality,* Volume 1, *An Introduction* (London: Allen Lane, 1979).
Freeman, M. and Spencer, S., 'Immigration Control, Black Workers and the Economy', *British Journal of Law and Society* (1979), 6: 1, 53–81.
Fryer, P., *Staying Power: the History of Black People in Britain* (London: Pluto, 1984).
Gainer, B., *Alien Invasion: the Origins of the Aliens Act of 1905* (London: Heinemann, 1972).
Galton, F., 'Hereditary Talent and Character', *Macmillan's Magazine* (1865), 12, 318–27.
Gartner, L., *History of the Jews in Modern Times* (Oxford: Oxford University Press, 2000).
Gates, R., *Heredity and Eugenics* (London: Constable, 1923).
Gates, R., *Human Genetics* (2 volumes) (New York: Macmillan, 1946).
Gates, R., *Pedigrees of Negro Families* (Philadelphia: Blackiston, 1949).
Gates, R., *The Emergence of Racial Genetics* (New York: International Association for the Advancement of Ethnology and Eugenics, 1960).
Gibbons, M. et al., *The New Production of Knowledge: the Dynamics of Science and Research in Contemporary Societies* (London: Sage, 1994).
Gilbert, M., *Exile and Return: the Emergence of Jewish Statehood* (London: Weidenfeld and Nicolson, 1978).
Gillman, P. and L., *'Collar the Lot!' – How Britain Interned and Expelled its Wartime Refugees* (London: Quartet Books, 1980).

Gilman, S., *The Jew's Body* (London: Routledge, 1991).
Gilroy, P., *Between Camps: Nations, Cultures and the Allure of Race* (London: Penguin, 2000).
Glass, R., *Newcomers: the West Indians in London* (London: Allen and Unwin, 1960).
Goldberg, D., *Racist Culture: Philosophy and the Politics of Meaning* (Oxford: Blackwell, 1993).
Golding, L., *Magnolia Street* (St Albans: Gainsborough, 1932).
Golding, L., *The Jewish Problem* (London: Penguin, 1938).
Goulbourne, H., *Ethnicity and Nationalism in Post Imperial Britain* (Cambridge: Cambridge University Press, 1991).
Gould, S.J., *The Mismeasure of Man* (London: Penguin, 1981).
Gregory, J., *The Menace of Colour* (London: Seeley Service, 1925).
Gregory, R., *Science in Chains* (London: Macmillan, 1941).
Grebenik, E., 'Demographic Research in Britain 1936–86', *Population Studies* (1991), 45, 3–30.
Haddon, A., *The Races of Man* (Cambridge: Cambridge University Press, 1924).
Haldane, J.B.S., *Callinicus: a Defence of Chemical Warfare* (London: Kegan Paul, 1925).
Haldane, J.B.S., *Heredity and Politics* (London: Allen and Unwin, 1938).
Haldane, J.B.S., *Science and Everyday Life* (Harmondsworth: Penguin, 1939).
Haldane, J.B.S., *Keeping Cool and Other Essays* (London: Chatto and Windus, 1940).
Haldane, J.B.S., *Science Advances* (London: Allen and Unwin, 1947).
Halsey, A., *A History of Sociology in Britain: Science, Literature and Society* (Oxford: Oxford University Press, 2004).
Hampshire, J., *Citizenship and Belonging: Immigration and the Politics of Demographic Governance in Postwar Britain* (Basingstoke: Palgrave Macmillan, 2005).
Hansen, R., *Citizenship and Immigration in Post War Britain: the Institutional Origins of a Multi-Cultural Nation* (Oxford: Oxford University Press, 2000).
Hasian Jr., M., The *Rhetoric of Eugenics in Anglo-American Thought* (Athens: University of Georgia Press, 1996).
Hawkins, M., *Social Darwinism in European and American Thought: Nature as Model and Nature as Threat 1860–1945* (Cambridge: Cambridge University Press, 1997).
Hayes, N. and Hill, J. (eds), *'Millions like us?': British Culture in the Second World War* (Liverpool: Liverpool University Press, 1999).
Hearnshaw, L., *Cyril Burt Psychologist* (London: Hodder and Stoughton, 1979).
Hickman, M., *Religion, Class and Identity: the State, the Catholic Church and the Education of the Irish in Britain* (Aldershot: Avebury, 1995).
Hoch, P., *White Hero Black Beast: Racism, Sexism and the Mask of Masculinity* (London: Pluto, 1979).
Hogben, L., *Genetic Principles in Medicine and Social Science* (London: Williams and Norgate, 1931).
Hogben, L., *Mathematics for the Million: a Popular Self Educator* (London: Allen and Unwin, 1936).
Hogben, L., *The Retreat from Reason* (London: Watts and Co, 1936).
Hogben, L., *Science for the Citizen: a Self Educator Based on the Social Background of Scientific Discovery* (London: Allen and Unwin, 1938).
Hogben, L., *Dangerous Thoughts* (London: Allen and Unwin, 1939).

Holmes, C., *Anti-Semitism in British Society 1876–1939* (London: Edward Arnold, 1979).

Holmes, C., 'Public Opinion in England and the Jews 1914–18', *Michael 10* (1986), 76–95.

Holmes, C., *John Bull's Island: Immigration and British Society 1871–1971* (Basingstoke: Macmillan, 1988).

Holmes, C., *A Tolerant Country? Immigrants, Refugees and Minorities in Britain* (London: Faber & Faber, 1991).

Hughes, A.G., 'Jews and Gentiles: Their Intellectual and Temperamental Differences', *Eugenics Review* (1928), 20: 2, 89–94.

Hutton, C., *Race and the Third Reich: Linguistics, Racial Anthropology and Genetics in the Dialectic of Volk* (Cambridge: Polity, 2005).

Huxley, J., *Africa View* (London: Chatto and Windus, 1931).

Huxley, J., 'Galton Lecture', *Eugenics Review* (1936), 28: 1, 11–31.

Huxley, J., 'Men of Science and the War', *Nature* (1940), 146: 3691, 107–8.

Huxley, J., *Argument of Blood* (London: Macmillan, 1941).

Huxley, J., 'Scientists in Uniform: England', *The New Republic*, 28/4/41, 590–1.

Huxley, J., *The Uniqueness of Man* (London: Chatto and Windus, 1941).

Huxley, J., *Soviet Genetics and World Science: Lysenko and the Meaning of Heredity* (London: Chatto and Windus, 1949).

Huxley, J., *Man's New Vision of Himself* (Natal: EP and Commercial Printing, 1960).

Huxley, J., 'Eugenics in Evolutionary Perspective', *Nature* (1962), 195: 4838, 227–8.

Huxley, J., 'Eugenics in Evolutionary Perspective', *Perspectives in Biology and Medicine* (1962), 6: 2, 155–87.

Huxley, J., 'JBS Haldane', *Encounter* (1965), 65: 4, 60.

Huxley, J., *Memories I* (London: Allen and Unwin, 1970).

Huxley, J., *Memories II* (London: Allen and Unwin, 1973).

Huxley, J. and Haddon, A.C., *We Europeans: a Survey of Racial Problems* (London: Jonathan Cape, 1935).

Isaac, J., *British Post-War Migration* (Cambridge: Cambridge University Press, 1954).

Jackson, J., 'The Scientific Attack on Brown v. Board of Education, 1954–1964', *American Psychologist* (2004), 59: 6, 530–7.

Jackson, W., *Gunnar Myrdal and America's Conscience: Social Engineering and Racial Liberalism 1938–87* (Chapel Hill: University of North Carolina Press, 1990).

James, R. (ed.), *Chips: the Diary of Sir Henry Channon* (London: Weidenfeld and Nicholson, 1967).

Jenkinson, J., 'The Black Community of Salford and Hull 1919–21', *Immigrants and Minorities* (1988), 7: 2, 166–83.

Jones, G., 'Eugenics and Social Policy Between the Wars', *The Historical Journal* (1982), 25: 3, 717–28.

Jones, G., *Science, Politics and the Cold War* (London and New York: Routledge, 1988).

Joravsky, D., *The Lysenko Affair* (Cambridge, Massachusetts: Harvard University Press, 1970).

Kadish, S., *Bolsheviks and British Jews: the Anglo-Jewish Community and the Russian Revolution* (London: Frank Cass, 1992).

Katagiri, Y., *The Mississippi State Sovereignty Commission: Civil Rights and States' Rights* (Jackson: University Press of Mississippi, 2001).

Kay, D. and Miles, R., *Refugees or Migrant Workers? European Volunteer Workers in Britain 1946–1951* (London: Routledge, 1992).

Keith, A., *The Antiquity of Man* (2 Volumes) (London: Williams and Norgate, 1915).

Keith, A., 'The Evolution of the Human Races', *Journal of the Royal Anthropological Institute of Great Britain and Ireland* (1925), 58: 306–21.

Keith, A., *The Place of Prejudice in Modern Civilisation: (Prejudice and Politics)* (London: Williams and Norgate, 1931).

Keith, A., *Essays on Human Evolution* (London: Watts and Co, 1946).

Kenny, M., 'Toward a Racial Abyss: Eugenics, Wickliffe Draper, and the Origins of the Pioneer Fund', *Journal of the History of the Behavioural Sciences* (2002), 38: 3, 259–83.

Kenny, M., 'Racial Science in Social Context: John R. Baker on Eugenics, Race, and the Public Role of the Scientist', *ISIS* (2004), 95: 394–419.

Kevles, D., *In the Name of Eugenics: Genetics and the Use of Human Heredity* (New York: Knopf, 1985).

King, C. and King, H., *'The Two Nations': the Life and Work of the Liverpool University Settlement and its Associated Institutions 1906–37* (London: Liverpool University Press, 1938).

King, R., *Race, Culture and the Intellectuals, 1940–1970* (Baltimore: Johns Hopkins University Press, 2004).

Klineberg, O. (ed.), *Characteristics of the American Negro* (New York: Harper, 1944).

Kochan, M., *Britain's Internees in the Second World War* (London and Basingstoke: Macmillan, 1983).

Kohn, M., *The Race Gallery: the Return of Racial Science* (London: Jonathan Cape, 1995).

Kohn, M., *A Reason for Everything: Natural Selection and the English Imagination* (London: Faber and Faber, 2005).

Knox, R., *The Races of Men* (London: Savill and Edwards, 1850).

Kushner, T., *The Persistence of Prejudice: Anti-Semitism in British Society during the Second World War* (Manchester and New York: Manchester University Press, 1989).

Kushner, T., *The Holocaust and the Liberal Imagination: a Social and Cultural History* (Oxford: Blackwell, 1994).

Kushner, T., *Remembering Refugees: Then and Now* (Manchester and New York: Manchester University Press, 2006).

Kushner, T. and Knox, K., *Refugees in an Age of Genocide: Global, National and Local Perspectives during the Twentieth Century* (London: Frank Cass, 1999).

Kushner, T. and Lunn, K. (eds), *The Politics of Marginality: Race, the Radical Right and Minorities in Twentieth Century Britain* (London: Frank Cass, 1990).

Lafitte, F., *The Internment of Aliens* (Harmondsworth: Penguin, 1940).

Lawson, S. (ed.), *To Secure These Rights: the Report of Harry S Truman's Committee on Civil Rights* (Boston and New York: Bedford/St Martins, 2004).

Layton Henry, Z., *The Politics of Race in Britain* (London: Social Science Research Board, 1980).

Levene, M., 'Going Against the Grain: Two Jewish Memoirs of War and Anti-War 1914–18', *Jewish Culture and History* (1999), 2: 2, 66–95.

Little, K., 'The Study of Racial Mixture in the British Commonwealth: Some Anthropological Preliminaries', *Eugenics Review* (1941), 32: 4, 114–20.
Little, K., 'Racial Mixture in Great Britain: Some Anthropological Characteristics of the Anglo-Negroid Cross', *Eugenics Review* (1942), 33: 4, 112–20.
Little, K., *Negroes in Britain: a Study of Race Relations in English Society* (London and Boston: Routledge and Kegan Paul, 1948).
Little, K., 'Behind the Colour Bar', *Current Affairs* (1950), 118: 1–19.
Logan, R., *Howard University: the First Hundred Years 1867–1967* (New York: New York University Press, 1969).
London, L., *Whitehall and the Jews, 1933–48: British Immigration Policy, Jewish Refugees and the Holocaust* (Cambridge: Cambridge University Press, 2000).
Lunn, K. (ed.), *Race and Labour in Twentieth Century Britain* (London: Frank Cass, 1985).
Lysenko, T. (trans. T. Dobzhansky), *Heredity and its Variability* (New York: Kings Crown Press, 1946).
McGucken, W., *Scientists, Society and the State: the Social Relations of Science Movement in Great Britain 1931–47* (Columbus: Ohio State University Press, 1984).
McGucken, W., 'The Social Relations of Science: the British Association for the Advancement of Science, 1931–1940', *Proclamations of the American Philosophical Society*, cxiii, 1994.
MacKenzie, D., 'Eugenics in Britain', *Social Studies of Science* (1976), 6: 3–4, 499–532.
MacKenzie, N. (ed.), Senior, C. and Manley, D., *The West Indian in Britain* (London: Fabian Colonial Bureau, 1956).
Macklin, G., *Very Deeply Dyed in Black: Sir Oswald Mosley and the Resurrection of British Fascism after 1945* (London and New York: IB Tauris, 2007).
MacLeod, R. and Collins, P. (eds), *The Parliament of Science: the British Association for the Advancement of Science, 1831–1981* (Northwood: Science Reviews, 1981).
Malik, K., *The Meaning of Race: Race, History and Culture in Western Society* (Basingstoke: Macmillan, 1996).
Margach, J., *The Anatomy of Power: an Enquiry into the Personality of Leadership* (London: W.H. Allen, 1979).
Marrus, M., *The Unwanted: European Refugees in the Twentieth Century* (Oxford and New York: Oxford University Press, 1985).
Mason, P. (ed.), *Man, Race and Darwin* (London: Oxford University Press, 1960).
Mason, P., *Race Relations: a Field of Study Comes of Age* (London: Lucas and Co, 1968).
Mazumdar, P., *Eugenics, Human Genetics and Human Failings: the Eugenics Society, its Sources and its Critics in Britain* (London and New York: Routledge, 1992).
Medvedev, Z., *The Rise and Fall of TD Lysenko* (New York: Columbia University Press, 1969).
Memmi, A., *Portrait of a Jew* (London: Eyre and Spottiswoode, 1963).
Miles, R., *Racism and Migrant Labour* (London: Routledge and Kegan Paul, 1982).
Miles, R., *Racism after 'Race' Relations* (London and New York: Routledge, 1993).
Montagu, A., *Man's Most Dangerous Myth: the Fallacy of Race* (New York: Columbia University Press, 1945).
Morant, G., *The Significance of Racial Differences* (Paris: UNESCO, 1952).
Moul, M. and Pearson, K., 'The Problem of Alien Immigration into Great Britain Illustrated by an Examination of Russian and Polish Alien Children', *Annals of Eugenics* (1925–6), 1, 6–127.

Mudge, G.P., 'The Menace to the English Race and to its Traditions of Present Day Immigration and Emigration', *Eugenics Review* (1920), 11: 4, 202–12.

Murphy, A., *From the Empire to the Rialto: Racism and Reaction in Liverpool 1918–48* (Birkenhead: Liver Press, 1995).

Myrdal, G., *An American Dilemma: the Negro Problem in Modern Democracy* (New York: Harper, 1944).

Needham, J., *The Nazi Attack on International Science* (London: Watts and Co, 1941).

Newby, I.A., *Challenge to the Court: Social Scientists and the Defense of Segregation 1954–1966* (Baton Rouge: Louisiana State University Press, 1967).

Nicholas, S., *The BBC, British Morale, and the Home Front War Effort 1939–45* (Oxford: Oxford University Press, 1992).

Nicholas, S., *The Echo of War: Home Front Propaganda and the Wartime BBC 1939–45* (Manchester and New York: Manchester University Press, 1996).

Nowotny, H., Scott, P. and Gibbons, M., *Re-Thinking Science: Knowledge and the Public in an Age of Uncertainty* (Cambridge: Polity, 2001).

Panayi, P., *Enemy in our Midst: Germans in Britain during the First World War* (Oxford: Berg, 1991).

Panayi, P., *Racial Violence in Britain 1840–1950* (Leicester: Leicester University Press, 1993).

Parkes, J., *An Enemy of the People: AntiSemitism* (Harmondsworth: Penguin, 1945).

Patterson, S., *Dark Strangers: a Study of West Indians in London* (London: Penguin, 1965).

Paul, K., *Whitewashing Britain* (Ithaca and London: Cornell University Press, 1997).

Peach, C., *West Indian Migration to Britain: a Social Geography* (London: Oxford University Press, 1968).

Pearson, K., *The Groundwork of Eugenics* (London: University College London Press, 1909).

Peel, R. (ed.), *Essays in the History of Eugenics* (London: The Galton Institute, 1997).

Phillips, T. and M., *Windrush: the Irresistible Rise of Multi-Racial Britain* (London: HarperCollins, 1998).

Pilkington, E., *Beyond the Mother Country: West Indians and the Notting Hill White Riots* (London: Tauris, 1988).

Porter, B., *The Refugee Question in Mid Victorian Politics* (Cambridge: Cambridge University Press, 1979).

Race and Culture (London: Royal Anthropological Institute and the Institute of Sociology, 1935).

Ragussis, M., *Figures of Conversion: 'the Jewish Question' and English National Identity* (Durham and London: Duke University Press, 1995).

Reynolds, D., 'The Churchill Government and the Black American Troops in Britain during World War Two', *Transactions of the Royal Historical Society* (1985), 5: 35, 113–33.

Reynolds, D., *Rich Relations: the American Occupation of Britain 1942–5* (London: HarperCollins, 1995).

Rich, P., 'Philanthropic Racism in Britain: the Liverpool University Settlement, the Anti-Slavery Society and the Issue of "Half-Caste" Children, 1919–51', *Immigrants and Minorities* (1984), 3: 1, 69–88.

Rich, P., *Race and Empire in British Politics* (Cambridge: Cambridge University Press, 1986).

Richards, G., *Race, Racism and Psychology: Towards a Reflexive History* (London and New York: Routledge, 1997).
Richmond, A., *Colour Prejudice in Britain: a Study of West Indian Workers in Liverpool 1941–51* (London: Routledge and Kegan Paul, 1954).
Richmond, A., *The Colour Problem* (Harmondsworth: Penguin, 1955).
Richmond, A., 'Sociological and Psychological Explanations of Racial Prejudice: Some Light on the Controversy from Recent Researches in Britain', *Pacific Sociological Review* (1961), 4: 2, 63–8.
Richmond, C., *Campaigner against AntiSemitism: the Reverend James Parkes, 1896–1981* (London: Vallentine Mitchell, 2005).
Rose, E. et al., *Colour and Citizenship: a Report on British Race Relations* (London: Oxford University Press, 1969).
Rose, H. and S., *Science and Society* (London: Allen Lane, 1969).
Rosenthal, M., *The Character Factory: Baden Powell and the Origins of the Boy Scout Movement* (New York: Pantheon, 1986).
Rossianov, K., 'Editing Nature: Joseph Stalin and the "New" Soviet Biology', *ISIS* (1993), 84: 4, 728–45.
Rumyaneck, J., 'The Comparative Psychology of Jews and Non-Jews: a Survey of the Literature', *The British Journal of Psychology* (1931), 11: 4, 409–23.
Saleeby, C., *The Eugenic Prospect: National and Racial* (London: Fisher Unwin, 1921).
Schaffer, G., '"Like a Baby with a Box of Matches": British Scientists and the Concept of "Race" in the Interwar Period', *British Journal for the History of Science* (2005), 38: 3, 307–24.
Schaffer, G., 'Re-Thinking the History of Blame: Britain and Minorities during the Second World War', *National Identities* (2006), 8: 4, 401–20.
Schaffer, G., '"Scientific" Racism Again?: Reginald Gates, the *Mankind Quarterly* and the Question of "Race" in Science after the Second World War', *Journal of American Studies* (2007), 41: 2, 253–78.
Searle, G., *The Quest for National Efficiency: a Study in British Politics and Political Thought 1899–1914* (Oxford: Blackwell, 1971).
Searle, G., *Eugenics and Politics in Britain 1900–14* (Leyden: Noordoff International Publishing, 1976).
Segev, T., *One Palestine, Complete: Jews and Arabs under the British Mandate* (New York: Metropolitan Books, 2000).
Sherwood, M., *Many Struggles: West Indian Workers and Service Personnel in Britain 1939–45* (London: Karia Press, 1985).
Sherwood, M., 'Lynching in Britain', *History Today* (1999), 49: 3, 21–3.
Sissons, M. and French, P., *The Age of Austerity* (Oxford: Oxford University Press, 1986).
Sivanandan, A., *A Different Hunger: Writings on Black Resistance* (London: Pluto Press, 1982).
Sleigh, M. and Sutcliffe, J. (eds), *The Origins and History of the Society for Experimental Biology* (London: The Society for Experimental Biology, 1966).
Smith, G., *When Jim Crow Met John Bull* (London: I.B. Tauris, 1987).
Smith, R., 'Biology and Values in Interwar Britain: CS Sherrington, Julian Huxley and the Vision of Progress', *Past and Present* (2003), 178, 210–42.
Solomos, J., *Race and Racism in Britain* (Basingstoke: Macmillan, 1989).

Soloway, R., *Demography and Degeneration: Eugenics and the Declining Birthrate in Twentieth Century Britain* (Chapel Hill and London: The University Press of North Carolina, 1990).
Southern, D., *Gunnar Myrdal and Black–White Relations* (Baton Rouge: Louisiana State University Press, 1987).
Soyfer, V. (trans. L. Gruliow and R. Gruliow), *Lysenko and the Tragedy of Soviet Science* (New Brunswick: Rutgers University Press, 1994).
Spencer, I., *British Immigration Policy since 1939* (London: Routledge, 1997).
Spier, E., *The Protecting Power* (London: Skeffington, 1951).
Stack, D., *The First Darwinian Left: Socialism and Darwinism 1859–1914* (Cheltenham: New Clarion Press, 2003).
Stadulis, E., 'The Resettlement of Displaced Persons in the United Kindgom', *Population Studies* (1952), 5: 3, 207–37.
Stent, R., *A Bespattered Page? The Internment of 'His Majesty's Most Loyal Enemy Aliens'* (London: André Deutsch, 1980).
Stepan, N., *The Idea of Race in Science: Great Britain 1800–1960* (London: Macmillan, 1982).
Stone, D., 'Race in British Eugenics', *European History Quarterly* (2001), 31: 3, 397–425.
Stone, D., *Breeding Superman: Nietzsche, Race and Eugenics in Edwardian and Interwar Britain* (Liverpool: Liverpool University Press, 2002).
Stone, D., 'Nazism as Modern Magic: Bronislaw Malinoswki's Political Anthropology', *History and Anthropology* (2003), 14: 3, 203–18.
Swift, R. and Gilley, S. (eds), *The Irish in the Victorian City* (London: Croom Helm, 1985).
Sword, K. et al., *The Formation of the Polish Community in Great Britain 1939–1950* (London: Caldra House, 1989).
Tabili, L., 'The Construction of Racial Difference in Twentieth Century Britain: the Special Restriction (Coloured Alien Seamen) Order, 1925', *Journal of British Studies* (1994), 33: 1, 54–98.
Tannahill, J., *European Volunteer Workers in Britain* (Manchester: Manchester University Press, 1958).
Thorne, C., *Border Crossings: Studies in International History* (Oxford: Basil Blackwood, 1988).
Trythall, A., 'The Downfall of Leslie Hore-Belisha', *Journal of Contemporary History* (1981) 16: 3, 391–408.
Tucker, W., *The Funding of Scientific Racism: Wickliffe Draper and the Pioneer Fund* (Urbana: University of Illinois Press, 2002).
UNESCO, *Statement on Race: an Annotated Elaboration and Exposition of the Four Statements on Race Issued by the UNESCO* (New York: OUP, 1972).
Wasserstein, B., *Britain and the Jews of Europe* (Oxford: Oxford University Press, 1988).
Waters, C. and Van Helden, A., *Julian Huxley: Biologist and Statesman of Science* (Houston: Rice University Press, 1992).
Watts, M., *The Jewish Legion and the First World War* (Basingstoke: Palgrave, 2004).
Wells, G.P., *Lancelot Thomas Hogben: Biographical Memoirs of Fellows of the Royal Society* (London: The Royal Society, 1978).

Wells, H.G., Wells, G.P., and Huxley, J., *The Science of Life: a Summary of Contemporary Knowledge about Life and Its Possibilities*, 3 vols (London: Amalgamated Press, 1929-30).

Werskey, G., *The Visible College: a Collective Biography of British Scientists and Socialists of the 1930s* (London: Allen Lane, 1978).

White, A., *The Modern Jew* (London: Heinemann, 1899).

White, A., *Efficiency and Empire* (London: Methuen, 1901).

Williams, V., *The Social Sciences and Theories of Race* (Urbana and Chicago: University of Illinois Press, 2006).

Wilmer, E. and Brunet, P., 'John Randal Baker, 1900-84', *Biographical Memoirs of the Fellows of the Royal Society* (1985), 31, 33-63.

Wilson, C., 'Racism and Private Assistance: the Support of West Indian and African Missions in Liverpool, England, during the Interwar Years', *African Studies Review* (1992), 35: 2, 55-76.

Zollschan, I., *Racialism Against Civilisation* (London: New Europe, 1942).

Index

Academic Assistance Council 84
Aikman, K.D. 47
Aliens Act 1905 11, 12, 18, 60
Aliens and Nationality Committee 14
Aliens Order 1920 13
Aliens Restriction Act 1914 12
Aliens Restriction Amendment Act 1919 12, 56
Anderson, John 83
Angell, Robert 120, 124
Anglo-Jewish Association 112
Anglo-Russian Military Service Agreement 55
Anti-Slavery Society 21
Apartheid 4, 70, 90, 106–7, 114, 119, 130, 132, 135, 137, 145–6, 149
Arandora Star 80
Argument of Blood 33–4, 75–9
Armstrong, Charles Wicksteed 47, 56

Baker, John Randal 100, 166–70
Banton, Michael, and the history of race 2, 6, 10, and UNESCO 125, post-war research 153, 155–6
Barkan, Elazar, and the relationship between science and society 3, and the decline of scientific racism 5, and the Geneticists' Manifesto 67, and the UNESCO Statements on race 123
Beamish, Tufton 112
Beddoe, John 27
Benedict, Ruth 73, 105, 153
Bernal, John Desmond 118
Bertram, G.C.L. 149–52
Beveridge, William 84
Bevin, Earnest 112
Bibby, Cyril 130–2
Biometrical research on race 27–8

Birth Control Investigation Committee (BCIC) 41
Black Britons, targeting of in immigration legislation 13, 148–65, attitudes to during the War 85–96, sexual fears about 157–60 (and see miscegenation), as disease carriers 88
Blacker, Carlos Paton 75, 125, and the Eugenics Society 40–1, 51, 98–102, 150, war service 63, post-war racial studies 137, 167, 169–70
Bland, Neville 81
Board of Deputies of British Jews 19, 83
Boas, Franz 72, 74, 105
Boer War, impact on racial thinking 11
Bolshevism, fear of immigrant involvement with 12, 13
Boresza, M. 118
Boyle, Edward 159
Brains Trust 65, 76
British Association for the Advancement of Science 48
British Council of Churches 158
British Social Hygiene Council 43
British Travel Certificates 162–3
Brown versus *Board of Education of Topeka* 73, 131, 141–2
Burt, Cyril 20
Butcher, John 56

Campbell, A.B. 65
Caradog Jones, David 17, 21, 23–4
Carnegie Institute 73
Carr Saunders, Alexander 75, and *We Europeans* 33–4
Carter, Bob 2
Central Office for Refugees 83
Cesarani, David 112–13

Index

Chamberlain, Houston Stewart 103, 168
Chamberlain, Neville 58
Cheyette, Bryan 53
Churchill, Winston 87
Clark, Kenneth B. 131, 141, 153
Cold War 115, 118–19, 123, 130, 134, 136
Comas, Juan 144
Commonwealth Immigrants 148–65
Commonwealth Immigrants Act (1962) 149, 156–7, 161, 164
Creech Jones, Arthur 94
Crew, Frank (FAE), founder of the Society of Experimental Biology 28, and the Eugenics Society 41, racial views of 47, 66, war role 63
Cummins, Howard 106
Current Anthropology 144

Daily Worker 49, 64
Darlington, Cyril Dean 126
Davies, Mary 16, 19
Davies and Hughes, report on the comparative intelligence of Jewish and non-Jewish children 16, 19, 20, 21
Degeneration, fears of 11
Distressed Relatives Scheme 112
Dover, Cedric 35, 39
Draper, Wickliffe 143, 151–2
Drummond Shields, T. 43
DuBois, William Edward Burghardt 103
Dunera 80
Dunn, Leslie Clarence 123, 128

Eagleston, A.J., report on 'alien immigration' 59–60
Eden, Anthony 94, 160
Einstein, Albert 118
Ellinger, U.H. 103, 105
Elliot Smith, Grafton 30–1
Éluard, Paul 118
Eugenics, movement in Britain 3
Eugenics Society, and its policies on race and immigration 8, 40–1, 51, 98–102, 149–52, origins of 9, intellectual coherence of 9, research into racial crossing 17
European Volunteer Workers 107–14
Evans-Gordon, William 11, 57

First World War, impact on immigration policy 12, 54–6
Firth, Raymond 29–30
Fisher, Ronald Aylmer 126
Fleming, Rachel 16, 22, 51
Fletcher, Muriel, Report on 'Colour Problem' in Liverpool 16, 21–5, 47, 51
Fleure, Herbert John 50, research into Anglo-Chinese children 16, 17, 22, and UNESCO 125
Foot, Paul 156, 159
Forster, Edward Morgan 76

Galton, Francis 97, 99
Galton Lecture (1936) 42–4, 48, 66, 150, (1945) 41, 51, 98–9, 100, (1962) 41, 136
Garrett, Henry 142–3
Gates, Reginald 8, and the Race and Culture conference 29, 31–2, anti-Semitism 31, 37, 56–7, 102, 104, 138 and *We Europeans* 36–7, racial views 47–8, 66, 96–8, 137–41, 169, and Julian Huxley 48, 136, at Howard University 103–7, relationship with segregationists and the *Mankind Quarterly* 142–8, and the Eugenics Society 151–2
Gayre, George Robert 140–8, 169
Geneticists' Manifesto (1939) 65–7
Gilroy, Paul 5
Ginsberg, Morris 33, 120, 124
Glass, Ruth 154–7, 160
Gobineau, Arthur de 103, 168
Goldberg, David Theo 3
Golding, Louis 62
Goodrich, Edwin Stephen 44
Greene, Graham 76–9
Gregory, John Walter 47
Grigg, James 87–9
Guest, Ernest 90

Haddon, Alfred Cort, and *We Europeans* 32–9, 45, 48, 68
Haldane, John Burdon Sanderson opposition to Nazi Germany 26, 27, 68, 70, racial theories of 26, 27, 39, 40, 44–5, 66, 67, 69–70, 71, 90–1, 92–3, 134–7, 146, founder of the Society of Experimental Biology 28, Race and Culture conference 29, 30, attitude towards eugenics 40–1, and public education 49, defence of Jewish refugees 61, 83, 92, war role 63–5, 75, opposition to segregation in Rhodesia 90, attitude to USSR 116–19, and UNESCO 123, 127, 129, 148, and *The Geography of Intellect* 146–7
Halifax, Lord, (Edward Wood) 93
Hankey, Maurice 59
Hansberry, William 103
Harris, John 21
Harrison, G. Ainsworth 152
Hill, Archibald Vivian 84
Hitler, Adolf 19, 32, 42, 45–7, 58–9, 63–4, 69, 71, 74–5, 79, 81, 83, 93, 95–100, 103, 134, 137–8, 147, 168
Hogben, Lancelot 8, criticism of biometrics 28, founder of the Society of Experimental Biology 28, racial theories of 39, 40, 66, 67, 71–2, 134–5, 137, 146, attitude towards eugenics 40–1, and *Science for the Citizen* 49, 50, and public education 50, war role 63–4, support for Jewish refugees 83, opposition to apartheid 90, attitude to USSR 116
Holocaust, impact on science 96
Honigmann, Hans 84–5
Hooton, Earnest Albert 103, 140
Howard University 103–7
Hughes, Arthur, research on Jewish immigrant children 16, 19, research on Jews and Gentiles 20, 21

Huxley, Julian 8, and Charles Singer 1, opposition to Nazism 26, 32, 41–2, 68, 70, racial theories of 26, 39, 41–5, 66, 67, 69–70, 74, 91–5, 134–7, 146, founder of the Society of Experimental Biology 28, and *We Europeans* 32–9, 45, attitude towards eugenics 40–1, 151, Galton Lecture (1936) 43–4, 48, on the need for scientific engagement with the public 48–50, war role 63–4, 75, and the Brains Trust 65, and *Argument of Blood* 75–9, support for refugees 84–5, opinions on Empire 91–3, support for black Britons 94–5, attitude to USSR 116, 118, and UNESCO 120, 123–7, and Cyril Bibby 132–4
Huxley, Thomas Henry 133

Institute of Race Relations 154
Institute of Sociology 29
International Association for the Advancement of Ethnology and Eugenics 143
Internment, of enemy 'aliens' 79–86
Irish immigrants 161–2

Jewish Chronicle 14, 55
Jewish Health Organisation of Great Britain 19, 20, 50
Jews, immigration from Eastern Europe 11, attempts to restrict immigration of 11, 12, 13, as undesirable immigrants 18, racial assessments of 17, 18, 19, refugees from Nazism 53–62, 79–86, portrayed as cowards 53–6, 59, 81–2, as conspirators 56–7, 81, as criminals 56, as disease carriers 56–7, 60, in the European Volunteer Worker programmes 110–14
Jews' Free School 17, 19
Joad, Cyril 65, 76
Joliot-Curie, Frédéric 118

Jones, Greta, and the Eugenics Society 9, and the Geneticists' Manifesto 67, and Cold War science 118, 123
Journal of Experimental Biology 28, 36, 47, 136

Keith, Arthur, racial theories of 45–7, 96–8, 137–40, 144, and anti-Semitism 138–9
Kevles, Daniel, and the Eugenics Society 40, and the UNESCO Statements on race 123
Kindertransport policy 61
King, Richard 5, 15
Klineberg, Otto 73, 98, 105, 123, 153
Knox, Robert 27
Kohn, Marek 5, 166
Kristallnacht, British reactions to 58, 60, 61
Kushner, Tony 58, 81, 111–12

Lawson, H. 11
Leger, Fernand 118
Levy, Hyman 118
Little, Kenneth 74, 126, 153–4
Liverpool Association for the Welfare of Coloured People 23
Liverpool Association for the Welfare of Half-Caste Children 17, 21, 50–1
Lloyd George, Gwilym 161
Ludovici, Anthony 47
Lysenko, Trofim 115–19

Macmillan, Harold 87
Macmillan (War Pamphlet Series) 76
Malan, Daniel 137
Malik, Kenan 3, 6
Man (Journal of the Royal Anthropological Institute) 125–7, 129
Mankind Quarterly 142–8
Manley, Norman 163
Marshall, Tom 133
Martial, Rene 104
Martin, Kingsley 118

Mason, Philip 154
Mazumdar, Pauline, on the decline of scientific racism 15, 25, on the goals of the Eugenics Society 51
McGucken, William 4
Mead, Margaret 105
Medawar, Peter 167
Mendel, Gregor 28
Mendelian scientists, approach to racial study 28
Mental racial difference, ideas of 7, 42–3, 68–73, 122–30, 133–5, 146–7, 150, 168
Metraux, Alfred 125–6, 129–32
Miles, Robert 1, 2
Milne, A.A. 76
Miscegenation, perceptions of the effects of 21–4, as explained in *We Europeans* 35–6, 38–9, progressive thinking on 45, conservative thinking on 47, 150–2, government concerns about 86–8, social concerns about 157–61
'Mode 2' Science 4
Montagu, Ashley 98, 105, 121, 123, 127, 129–30
Moul, Margaret 16, 17, 18
Moul and Pearson, report on alien immigration 16, 17, 18, 20, 21, 54, 56, 60 (and criticism of) 19
Mourant, Arthur 127, 129
Mudge, G.P. 54
Muller, Hermann Joseph 44, 123, 136
Myrdal, Gunnar 73, 74, 98, 153

National Association for the Advancement of Coloured People 131
National Union of Students, Arthur Keith's lecture to 46
Nazism, impact on British science 26, 29, 30, 32, 36, 38, 39, 41–2, 46–7, 62, 65, 68, 75–9, 93–6, 98–100, 115, 120–1
Needham, Joseph 66, 68, 71–3, 75, 84
Newby, I. 141

Newsom, F.A. 84
Notting Hill riots 156, 159, 164

Parkes, James 84
Parkes Weber, Frederick 54–7
Patterson, Sheila 153, 155, 160
Pearson, Karl, research on immigrant Jewish children 16, 17, 18, and biometric research 27, 28
Penguin, Scientific publishing 48
Physical anthropology and racial study 27, 28
Physical racial difference, ideas of 7
Picasso, Pablo 118
Pitt-Rivers, George 31–2, 47
Profumo, John 163
Purves, D. 144–5

Race and Culture conference (1934) 29–32
Race riots (1919) 21
Racial difference, impact of perceptions of on immigration policy 11–14
Racial science, historiography of 5, decline of 5
Ramos, Arthur 120, 124
Rathbone, Eleanor 82
Reed, Douglas 82
Richmond, Anthony 87, 153–4, 159–60
Routh, Denis 94
Roxby, Percy 22, 23
Royal Anthropological Institute 29
Royal Commission on Alien Immigration 57
Royal Commission on Population 100–2, and European Volunteer Workers 108–9
Royal Society 50, 84

Saleeby, Caleb 9
San Francisco Examiner, publication of Huxley's 'Aryan Racial Myth Exploded' 42
Sargant Florence, Paul 152
Scientists and society, different understandings of race 2, inform each other on race 2, 3, 4, 10, 51–2, shared beliefs on race 79, 95–6
Segregation (USA) 4, 5, 7, 9, 70, 73, 90, 106–7, 119, 130–2, 137, 141–6, 149, 151, 153, 169
Seligman, Charles 30, and *We Europeans* 32, 33, 37
Semitic discourse 53, 57, 59, 111, 113
Singer, Charles, and Julian Huxley 1, 50, and *We Europeans* 32, 33–4, 37, and *Argument of Blood* 76–9, support for Jewish refugees 83
Smith, Roger 4, 9–10, 48, 51–2
Society for Experimental Biology 28
Society for the Protection of Science and Learning 84
Sociology, racial studies in 72–3
Soloway, Richard, and the Eugenics Society 40
Special Restriction (Coloured Alien Seamen) Order 1925 13, 14, 22
Stalin, Joseph 115–17
Stanley, Oliver 94
Stepan, Nancy, on the decline of scientific racism 3, 5, 15, 26
Stone, Dan 8, 26, 51
Szilard, Leo 84

Tate, Mavis 82
Taylor, A.J.P. 118
Tirala, Lothar 77
Torres-Bodet, Jaime 120, 126
Trevor, John B. 152
Trevor, J.C. 127, 129
Truman, Harry, Commission on Civil Rights 73, 131–2
Tucker, William 5
Twitchen, Henry 152
Twyman, F. 69
Tyrell, Francis 57

UNESCO, Statements and research on race 120–32, 137, 147–50
University of Aberystwyth 17
University of Liverpool 16, 17
USSR, science in the post-war period 115–19

Venereal disease, (and black troops) 88
Volksdeutsche, in the European Volunteer Worker programmes 110–14

Watt, Ian 163
We Europeans, and the idea of race 2, 3, 35–6, the writing of 32–4, as a challenge to Nazism 32, 35–6, 38, 49, as a limited challenge to the idea of race 38–9, influence on later race writing 67–8, 76–7, 96, 134
Wedgwood, Josiah Clement 81–2
Werskey, Gary, *The Visible College* 8, and scientific engagement in the war effort 67, and scientists during the Cold War 119, 170
West Indian Regiment 89–90
Weyher, Harry 151
White, Arnold 9, 53
Wild, Ernst 56
Winterton, Edward 81, 111
Wooton, Charles 21
Work, Munro 103
Wroclaw, conference in 118

Zollschan, Ignaz 74
Zuckerman, Solly 126–7, 129